The Great Silence

The Great Silence

John T. Prorock

First Printing, 2026

APA 7

Prorock, J. T. (2026). *The Great Silence - A Journey Through the Fermi Paradox*. Husky Publishing Company.

MLA 9

Prorock, John T. *The Great Silence - A Journey Through the Fermi Paradox*. Husky Publishing Company, 2026.

Chicago

Prorock, John T. *The Great Silence - A Journey Through the Fermi Paradox*. Philadelphia, PA: Husky Publishing Company, 2026.

ISBN/SKU: 979-8-218-91112-6
EISBN: 979-8-218-91113-3

I lovingly dedicate this book to my wife Gabby and daughter Aurora. In the infinite vastness of the cosmos, my girls are my lights in the darkness.

CONTENTS

Introduction: Where is Everybody?

The air in the Fuller Lodge at Los Alamos was thick with the casual chatter of brilliant minds. It was the summer of 1950, a time of burgeoning atomic optimism and Cold War anxiety. At one table, a group of physicists, including the celebrated Enrico Fermi, were enjoying a lighthearted lunch, their conversation drifting from recent UFO reports to the feasibility of faster-than-light travel. The topic was amusing, a fanciful break from the serious work of nuclear physics.

After a few moments of calculation, Fermi, known for his ability to estimate almost anything with surprising accuracy, concluded that our galaxy should be teeming with advanced civilizations. He reasoned that our sun is a relatively young star in a galaxy billions of years old. Countless other stars would have formed earlier, orbited by countless Earth-like planets. On some of these, life would have arisen, and on a fraction of those, intelligence would have evolved. Given the sheer timescale involved, some of these civilizations would have had a head start of millions, if not billions, of years. Even at slow, sub-light speeds, such a civilization would have had ample time to colonize the entire Milky Way.

The table fell quiet as the implications of his logic sank in. Fermi let the silence hang for a moment before asking a simple, four-word question that would echo through the decades and become one of the most profound and vexing questions in all of science: "But where is everybody?"

This is the Fermi Paradox. It is the stark, bewildering contradiction between the high probability of extraterrestrial intelligence and the complete, utter lack of any evidence for it. It is the Great Silence.

Our journey to understand this silence begins with a sense of cosmic vertigo. Look up at the night sky. The faint, milky band stretching across it is our home galaxy, a swirling disc of some 200 to 400 billion stars. For centuries, we believed these were mere points of light. Now we know better. Modern astronomy has revealed that nearly every one of those stars is a sun, and that most of them host their own families of planets. The Kepler Space Telescope alone has confirmed thousands of exoplanets and suggests there are more planets than stars in our galaxy.

To grasp these numbers is to teeter on the edge of comprehension. If you were to take all the grains of sand from every beach and desert on Earth, you would have a rough approximation of the number of stars in the observable universe. And orbiting these stars are potentially billions of worlds within the "habitable zone," that temperate sliver of space where conditions might be just right for liquid water to pool on a rocky surface—the one ingredient we deem essential for life as we know it.

Given these staggering odds, our universe should be a bustling cosmic metropolis, a vibrant tapestry of life, intelligence, and culture. We should see the grand engineering projects of Type III civilizations harnessing the energy of entire galaxies. We should hear the chatter of a billion worlds on the cosmic radio waves. We should have found artifacts, probes, or at least the chemical signatures of alien biospheres in the starlight. Instead, we have found nothing. The heavens are profoundly, unnervingly quiet. But why?

This book is an investigation into that quiet. We will embark on a journey that will take us from the microscopic definition of life to the mind-bending possibilities of interdimensional physics. In Part I, we will begin by defining our terms: before we can find life, we must first establish clear scientific parameters for what exactly we consider to be "alive." We will explore the consensus model of biology and then immediately challenge it, peering into the gray areas where the line between living and non-living blurs.

In Part II, we will quantify the mystery, using Dr. Frank Drake's famous equation as our guide. We will break down each variable, from the rate of star formation to the probable lifespan of a technological society, replacing speculation with the latest scientific data where possible, and confronting the vast unknowns that remain.

Part III forms the heart of our inquiry, a comprehensive exploration of the proposed solutions to the Great Silence. We will weigh the possibility that Earth is a cosmic fluke, a "Rare Earth" in a sterile universe. We will confront the terrifying idea of a "Great Filter" that dooms most emerging life, and the paranoid logic of the "Dark Forest," where silence is the only strategy for survival. We will entertain the notion that we are being deliberately watched in a galactic "Zoo," or that our entire reality is nothing more than a sophisticated computer simulation.

Finally, in Part IV, we will look to the future, examining the next generation of telescopes and projects that may, at long last, break the silence. Throughout this journey, we will find that the Fermi Paradox acts as a grand mirror. Each potential solution reflects our own civilization's hopes, fears, and ultimate limitations. The question, "Where is everybody?" inevitably forces us to ask another, even more critical question: "Who are we, and where are we going?"

The silence of the universe is a challenge. It beckons us to listen more closely, to think more deeply, and to explore more boldly. Let us begin.

1

Chapter 1: The Consensus on Life

When we speak of a search for extraterrestrial "life," we invoke a word that is at once intimately familiar and notoriously difficult to define. We intuitively recognize it in a soaring bird, a blooming flower, or a wriggling bacterium. But to build a scientific search, intuition is not enough. We need a testable, empirical framework—a checklist. What exactly are the fundamental properties that separate a living organism from an inert rock or a complex, but non-living, chemical reaction?

While debates in biology continue to refine the edges of the definition, a broad consensus has formed around a series of core characteristics. These are not arbitrary rules but observed properties that life on Earth—the only life we know—universally employs to survive, thrive, and perpetuate itself. To be considered "life" as we currently understand it, an entity must possess most, if not all, of the following nine characteristics:

1. **Cellular Organization**
   The foundational principle of known biology is that all living things are composed of one or more cells. The cell is the basic structural, functional, and biological unit of all known organisms. It is the vessel of life. A single-celled bacterium is an organism in its own right, a self-contained marvel of complexity. A

human being is a cooperative of trillions of cells, each specialized for a different task. The critical element is the cell membrane, a fatty barrier that separates the intricate, orderly chemical ballet inside the cell from the chaos of the outside environment. Without this fundamental division, the processes of life would dissipate into the surroundings. When we search for life, we are, in our most basic model, searching for evidence of these self-contained biological units.

2. **Homeostasis**

   Life is a precarious balancing act. To survive, an organism must maintain a stable internal environment, even when the external world is in constant flux. This ability is called homeostasis. Consider the human body: whether you are in a sweltering desert or a frozen tundra, your internal body temperature remains stubbornly fixed around 98.6°F (37°C). This is achieved through a complex system of feedback loops. If you get too hot, you sweat to cool down; if you get too cold, you shiver to generate heat. The same principle applies to blood pH, sugar levels, and water content. This constant internal regulation is a hallmark of life, a ceaseless effort to maintain order in the face of entropy.

3. **Respiration**

   In a biological context, respiration is more than just breathing. It is the set of metabolic reactions that convert chemical energy from nutrients into adenosine triphosphate (ATP), the universal energy currency of the cell. For animals, this typically involves taking in oxygen and releasing carbon dioxide. For most terrestrial plant life, the process is reversed during photosynthesis, absorbing carbon dioxide and releasing oxygen. This exchange of gases facilitates the chemical reactions that power the organism, making it an observable and critical function of being alive.

4. **Metabolization**

   Metabolism is the sum of all chemical processes that occur within an organism to sustain life. It is the engine of being. This encompasses both catabolism, the breaking down of complex molecules to release energy (like digestion), and anabolism, the using of that energy to build complex components like proteins and nucleic acids. For an animal, metabolization includes eating, drinking, and digesting. For a plant, it is the elegant process of photosynthesis, converting sunlight, water, and carbon dioxide into glucose—a form of chemical energy. Any living entity must acquire and use energy to maintain its structure, grow, and reproduce.

5. **Excretion**

   Every chemical factory produces waste, and the factory of life is no exception. Excretion is the process by which living organisms eliminate the waste products of metabolism and other non-useful materials. In humans, this includes exhaling carbon dioxide, sweating, and the urinary system filtering waste from the blood. Without an effective means of waste removal, toxic substances would accumulate and quickly prove fatal. This process of purification is essential for maintaining the delicate internal balance of homeostasis.

6. **Growth**

   Living things do not remain static; they grow. This is not simply an increase in mass, but a process of development and increasing complexity, guided by a genetic blueprint. A single fertilized egg grows into a fully formed adult. A tiny seed grows into a towering sequoia. This growth is an irreversible increase in the size of an organism, a progression through life stages, each more complex than the last.

7. **Movement**

   All living beings move in some way. This may be obvious locomotion, like a cheetah sprinting across the savanna, but it can also be far more subtle. Plants exhibit movement as they turn their leaves toward the sun (phototropism) or as their roots navigate through the soil. Even within a seemingly stationary organism, there is constant internal movement: the transport of nutrients, the flow of cytoplasm within a cell, the beating of a heart. Movement, whether external or internal, is a fundamental sign of an active, living system.

8. **Reproduction**

   Perhaps the most profound characteristic of life is its ability to create more of itself. Reproduction is the biological process by which new individual organisms—"offspring"—are produced from their "parents." This can be as simple as a bacterium splitting in two (asexual reproduction) or as complex as the courtship and mating rituals of mammals (sexual reproduction). While an individual organism can live without reproducing, reproduction is absolutely essential for the continuation of a species. It is life's strategy for achieving a form of immortality.

9. **Response to Stimuli**

   Living organisms are not passive passengers in their environment; they actively interact with it. The ability to detect and respond to external stimuli is crucial for survival. A plant grows towards a source of light, a bacterium swims away from a toxic chemical, and an animal flees at the sound of a predator. These responses, known as taxis or tropisms, allow an organism to adapt its behavior and physiology to changing conditions, find resources, and avoid danger.

Together, these nine characteristics form the scientific bedrock for our search for life. They provide a template, a set of observable phenomena that, if detected on a distant world, could be the first credible evidence that we are not alone. Yet, as we shall see, this template is fragile, and its edges are frayed with exceptions and paradoxes that force us to question the very definition of being alive.

2

Chapter 2: The Edge of Life

The nine characteristics outlined in the previous chapter provide a robust and useful model for identifying life. They serve as our scientific baseline, the foundation upon which multi-billion dollar telescopes and interstellar probes are designed. Yet, science is a process of constant refinement, and our definition of life is far from absolute. Nature, in its boundless creativity, delights in producing exceptions. To fully appreciate the challenge of searching for life elsewhere, we must first explore the strange and blurry boundary between the living and the non-living here on Earth.

This exploration begins with a simple, stubborn animal: the mule. A hybrid offspring of a female horse and a male donkey, the mule is a paragon of strength and endurance. It possesses a cellular structure, maintains homeostasis, metabolizes food, grows from a foal to an adult, and responds to stimuli with a famously strong will. By almost every measure, it is vibrantly alive. Yet it fails one of our most fundamental tests: reproduction. Mules are, with exceptionally rare exceptions, sterile. If the ability to create new life of the same kind is a prerequisite for being a living organism, then one could be forced into the absurd argument that the mule is not a living being.

Of course, no biologist would make this argument. We understand that the mule is an individual member of a sterile hybrid species. It is alive, but it represents a dead end on the evolutionary tree. The mule paradox highlights a critical distinction: are we defining the life of an individual or the life of a species? For a species to be considered "living," it must be able to perpetuate itself. For an individual, however, the inability to reproduce does not negate its own life. This nuance is crucial. When we search for life on other planets, are we looking for a single, sterile organism, or are we looking for a self-perpetuating biosphere?

The challenge deepens when we consider non-biological phenomena that mimic the characteristics of life with uncanny accuracy. Consider fire. A flame can be said to pass many of our tests. It moves, dancing and licking at its fuel source. It grows, starting from a small spark and consuming an entire forest. It metabolizes fuel and respires by consuming oxygen while excreting carbon dioxide and smoke. It maintains homeostasis in a way, sustaining a relatively constant high temperature at its core. It can even be said to reproduce, as sparks from a forest fire can be carried by the wind to start new fires miles away. And it certainly responds to stimuli, flaring up with a gust of wind or dying down when doused with water.

Why, then, is a mule that fails one test considered alive, while a flame that passes nearly all of them is not? The answer lies in the characteristics that fire lacks: cellular organization and, most importantly, heredity. Fire has no genetic blueprint. Its "reproduction" is simply a physical chain reaction, not a biological process guided by DNA. Its complexity is emergent, but it does not evolve. There is no lineage, no passing of traits, no potential for adaptation over generations. Fire is a process, but it is not an organism.

This distinction brings us to the true razor's edge of biology, to entities that make even seasoned scientists question their definitions. Consider the virus. Is it alive? A virus is a packet of genetic material (DNA or RNA) encased in a protein shell. It is a masterpiece of minimalist design. Outside of a host cell, it is as inert as a grain of salt. It does not metabolize, it does not grow, and it cannot reproduce. But once it finds a host, it springs into a terrifying semblance of life. It injects its genetic code, hijacks the cell's machinery, and forces it to create thousands of new copies of the virus. In a sense, it outsources the characteristics of life. It possesses heredity and evolves through natural selection with terrifying speed, but it cannot perform the basic functions of life on its own. It exists in a quantum-like state of being, simultaneously living and non-living.

Pushing even further into this strange realm, we find prions. These are not viruses or bacteria; they are misfolded proteins. They contain no genetic material at all. Yet, they can "reproduce" in a horrifying way. When a prion encounters a normally folded version of the same protein, it induces that protein to misfold into the prion's shape. This sets off a chain reaction that can lead to devastating neurodegenerative diseases like Mad Cow Disease. Prions spread and cause damage, but they lack almost every single one of our nine characteristics, save for a perverse form of reproduction. Are they alive? Almost no one would say so. But they demonstrate that even complex, self-propagating behavior can arise from systems we would classify as non-living.

These edge cases—the sterile mule, the life-mimicking fire, the zombie-like virus, and the shape-shifting prion—are not mere curiosities. They are essential lessons. They teach us that our Earth-centric checklist, while useful, is likely incomplete and perhaps even naïve. The universe may have devised forms of life based on principles we have yet to imagine, life that might fail many, or even all, of our tests while still being unambiguously alive. As we prepare our instruments and aim them at the stars, we must carry with us a profound sense of humility, ready to have our most fundamental assumptions about life shattered.

How can we effectively search for "life" throughout the cosmos if we are unable to perfectly define what exactly constitutes life within our own planet's biosphere? If we can't thoroughly understand life on Earth, then it will certainly prove arduous if not impossible to comprehend life that is silicon-based, breathes methane, lives in cyanide, or is of another dimension entirely. An expanded mentality and more open-minded approach will be paramount to humanity's efforts in this endeavor. The search for "life as we know it" may be the only practical place to start, but the Great Silence may be telling us that we need to start looking for life as we don't know it.

3

Chapter 3: The Spark of Abiogenesis

How does life come from non-life? Short answer: it doesn't. Long answer: we don't know, but there are theories As far as modern science can explain, life is a process of biology, but it must have begun as a process of chemistry. Every organism on Earth, from the blue whale to the bacterium on your skin, can trace its lineage back to a single, common ancestor. But what came before that ancestor? How did the universe bridge the seemingly infinite gulf between inert, non-living matter—minerals, water, and gases—and the first, self-replicating, living cell? This is the question of abiogenesis, the genesis of life from non-life. It is arguably the most profound mystery in all of science and a prime suspect in our investigation into the Great Silence.

The transition from a mere collection of organic molecules to a structured, metabolizing, and reproducing entity is a jump of such staggering complexity that it could very well be the single greatest barrier to life in the cosmos. If this step is common, we should expect the universe to be teeming with at least simple life. But if it is an event of infinitesimal probability, a chemical miracle, then the silence we perceive may be the natural state of a universe where life is the rarest of exceptions. To understand the odds, we must first explore how this miracle might have happened here.

The "Warm Little Pond" and the Primordial Soup

The first coherent scientific hypothesis for abiogenesis began not in a laboratory, but in the mind of Charles Darwin. In an 1871 letter, he mused about the possibility of life originating in some "warm little pond, with all sorts of ammonia and phosphoric salts, light, heat, electricity, etc., present." He imagined that a protein compound could have been chemically formed, ready to undergo still more complex changes. This was a revolutionary thought: life was not a mystical force, but a chemical process that could be explained by natural laws.

This idea lay dormant until the 1920s, when Russian biochemist Alexander Oparin and British scientist J.B.S. Haldane independently revived it. They theorized that the early Earth's atmosphere was vastly different from today's. It was a "reducing" atmosphere, rich in gases like methane, ammonia, and hydrogen, with very little free oxygen. Oxygen, while essential for us, is a highly reactive element that tends to break down complex organic molecules. Its absence on the early Earth would have allowed these molecules to persist and accumulate.

Oparin and Haldane proposed that with energy sources like powerful lightning storms and intense ultraviolet radiation from the young sun, the simple inorganic gases in the atmosphere could have been synthesized into simple organic molecules, such as amino acids and nucleotides—the building blocks of proteins and DNA. These molecules would have rained down into the oceans, which slowly became a vast, dilute "primordial soup" of organic compounds. In this chemical broth, over millions of years, these simple molecules could have linked together to form more complex polymers, eventually leading to the first living cells.

For decades, this remained an elegant but untested hypothesis. Then, in 1952, a young graduate student named Stanley Miller, working under the guidance of Nobel laureate Harold Urey at the University of Chicago, decided to put the primordial soup to the test. The Miller-Urey experiment was as simple as it was brilliant. They created a closed, sterile glass apparatus containing water to represent the ocean and a mixture of gases—methane, ammonia, and hydrogen—to simulate the early atmosphere. They heated the water to create vapor and passed it through the gases, then subjected the mixture to repeated electrical sparks to simulate lightning. A condenser cooled the "atmosphere," causing the water to condense and rain back into the "ocean," carrying with it any new molecules that had formed.

They let the experiment run for a week. The initially clear water began to turn a murky pink, then a deep, cloudy red. When they analyzed the contents of the flask, their findings were electrifying. The water was now rich with organic compounds, including several types of amino acids, the fundamental building blocks of the proteins that drive all of life's essential functions.

The significance of the Miller-Urey experiment cannot be overstated. It demonstrated, for the first time, that the transition from simple inorganic matter to the complex organic building blocks of life was not a near-impossible leap, but a predictable and perhaps even easy step under the right conditions. The essential components for life, it seemed, could cook themselves.

The Deep Sea Forge: Hydrothermal Vents

For all its success, the primordial soup model has faced challenges. The surface of the early Earth was an exceptionally violent place. It was subject to a relentless bombardment by asteroids and comets, and with no significant ozone layer, it was bathed in sterilizing levels of ultraviolet radiation. Some scientists argue that any organic molecules forming in a surface-level "warm little pond" would have been destroyed before they could develop further. This has led to an alternative and compelling hypothesis: life did not begin in the sunlit shallows, but in the crushing darkness of the deep ocean.

In 1977, scientists exploring the Galápagos Rift discovered a phenomenon that would change our understanding of biology: hydrothermal vents. These are fissures on the seafloor, often called "black smokers," that spew superheated, mineral-rich water from the Earth's interior. Far from being barren, these vents were surrounded by thriving, exotic ecosystems, teeming with giant tube worms, crabs, and shrimp—all powered not by sunlight, but by chemical energy from the vents in a process called chemosynthesis.

Scientists soon realized that these vents offered a perfect, protected cradle for the origin of life. They provided:

1. **Protection:** The deep ocean would have shielded fragile, emerging life from the harsh UV radiation and frequent impacts on the surface.
2. **Chemical Energy:** The steep gradients in chemistry and temperature between the hot, alkaline vent fluid and the cooler, more acidic ocean water create a natural energy source, a form of chemiosmotic potential that is remarkably similar to the way our own cells generate energy.
3. **Rich Raw Materials:** The vents spew a cocktail of chemicals essential for life, including hydrogen sulfide, iron, and manganese.
4. **Natural Compartments:** The porous, iron-sulfide rock structures around the vents are riddled with tiny, interconnected pockets. These pockets could have acted as natural "cells," concentrating the necessary chemicals and serving as incubators for the first metabolic reactions, long before the evolution of the cell membrane. This is the foundation of the "iron-sulfur world" hypothesis.

In this view, life began not as a soup, but as a deep-sea forge, a self-sustaining metabolic process clinging to the mineral surfaces of a hydrothermal vent, only later developing the genetic code and cellular membranes needed to venture out into the open ocean.

Delivery from the Cosmos: Panspermia

There is a third, more provocative possibility: what if life did not begin on Earth at all? The theory of Panspermia (from the Greek for "seeds everywhere") proposes that the basic building blocks of life, or even primitive microbial life itself, are common in the universe and were delivered to our planet from space.

This idea, once relegated to science fiction, has gained credibility from two key lines of evidence. First, we know that some terrestrial microbes are astonishingly resilient. Organisms like tardigrades ("water bears") and certain bacteria can survive the extreme conditions of space: near-total vacuum, freezing temperatures, and intense radiation. This raises the possibility that microbes could survive a journey through space, perhaps encased within a rock that was blasted off a life-bearing planet by an asteroid impact.

Second, we have direct proof that the building blocks of life are formed in space. The Murchison meteorite, which fell in Australia in 1969, was found to contain over 90 different amino acids, many of which are not found in terrestrial life. This confirms that the chemistry to create life's components is not unique to Earth; it occurs naturally in asteroid belts and interstellar clouds.

It is important to note that Panspermia does not solve the problem of abiogenesis. It simply moves the location. If life arrived here from Mars, we must then ask how life began on Mars. However, it has profound implications for the Fermi Paradox. If life can spread between planets and even star systems, then a single successful origin event could be enough to "seed" an entire region of the galaxy.

Whether in a sun-drenched pond, a dark ocean abyss, or the heart of a distant nebula, the journey from chemistry to biology had to occur at least once. The immense, unanswered question is how large the gap truly is between a collection of amino acids in a flask and a fully-fledged, self-replicating cell with a stable genetic code. That chasm—the true origin of life—remains shrouded in mystery. It may well be the First Great Filter, the first and perhaps the highest hurdle that life must overcome. If the spark of abiogenesis is a one-in-a-trillion event, then the Great Silence is not a paradox; it is the expected norm.

4

Chapter 4: A History of the Search

Before the Fermi Paradox became a formalized problem, it existed as an intuitive dream. The notion that we might communicate with other worlds is an ancient one, but for most of human history, it remained in the realm of philosophy and fantasy. It wasn't until the 20th century that science provided a tool that could turn this dream into an actual experiment: the radio telescope. With the ability to detect electromagnetic radiation far beyond the spectrum of visible light, humanity, for the first time, had a listening device capable of eavesdropping on the cosmos. The Search for Extraterrestrial Intelligence, or SETI, had begun. This chapter is the story of that search—a story of quiet optimism, shoestring budgets, ingenious engineering, and the profound, deeply human desire to know if we are alone.

The Birth of an Idea

The scientific foundation for SETI was laid in 1959. In a landmark paper published in the journal Nature titled "Searching for Interstellar Communications," physicists Philip Morrison and Giuseppe Cocconi made a startlingly logical proposal. They argued that if other intelligent civilizations existed and wished to communicate, they would likely choose a method that was efficient and could travel vast interstellar distances. The obvious candidate was radio waves.

But which frequency would they choose? The radio spectrum is immense. Searching it all would be an impossible task. Morrison and Cocconi reasoned that an advanced civilization would select a non-arbitrary, universally significant frequency—a cosmic "hitching post." They proposed the 1,420-megahertz (21-centimeter wavelength) line, the natural emission frequency of neutral hydrogen. Hydrogen is the most abundant element in the universe, so any civilization with even a rudimentary knowledge of radio astronomy would know of this frequency. It was a universal beacon, the "murmur of the universe," and the perfect place to listen for a deliberate signal. Their paper concluded with a powerful call to action: "The probability of success is difficult to estimate; but if we never search, the chance of success is zero."

Project Ozma: The First Step

One man who needed no convincing was a young radio astronomer named Frank Drake. Independently, Drake had reached the same conclusions as Morrison and Cocconi. Armed with their paper as justification, he managed to secure a modest amount of funding for a pioneering experiment he dubbed Project Ozma, named after the princess from L. Frank Baum's Oz books—a ruler of a "distant land populated by strange and exotic beings."

In the spring of 1960, at the National Radio Astronomy Observatory in Green Bank, West Virginia, Drake pointed the observatory's 85-foot radio telescope toward two nearby, sun-like stars: Tau Ceti and Epsilon Eridani. For several weeks, he and his small team meticulously scanned the 21-centimeter hydrogen line, listening for any signal that was structured, pulsed, or in any way artificial.

The experiment was fraught with challenges. The equipment was primitive by today's standards, and the team was constantly plagued by terrestrial interference from secret military experiments and other sources. On the very first day of observation, the chart recorder suddenly registered a powerful, pulsing signal. For a heart-stopping moment, the team believed they had made the most important discovery in human history. The excitement was short-lived; they soon determined the signal was coming not from Epsilon Eridani, but from a secret high-altitude military aircraft.

Project Ozma ultimately detected no extraterrestrial signals. But its importance was monumental. It established that the search for alien intelligence was a valid field of scientific inquiry. It was a transition from "if" to "how." Drake had taken the first, tentative step into the Great Silence, and in doing so, he had laid the groundwork for everything that would follow.

The Arecibo Message: A Shout into the Void

For years, SETI remained a listening exercise. But what if we wanted to send a message of our own? In 1974, to celebrate a major upgrade to the colossal Arecibo Radio Telescope in Puerto Rico—then the largest in the world—Frank Drake and Carl Sagan conceived of a powerful "message in a bottle" to broadcast to the stars.

The Arecibo Message was a masterpiece of information theory. It was a simple pictogram, encoded in binary, consisting of 1,679 bits. This number was chosen because it is a semiprime (the product of two prime numbers, 23 and 73), suggesting to any mathematically inclined recipient that the data should be arranged in a rectangular grid of 23 columns by 73 rows.

When arranged correctly, the graphic revealed a wealth of information about humanity and our place in the cosmos. It depicted:

- The numbers one through ten in binary.
- The atomic numbers of the key elements for life: hydrogen, carbon, nitrogen, oxygen, and phosphorus.
- The chemical formulas for the sugars and bases in DNA.
- A graphical representation of the DNA double helix.
- A simple figure of a human being, with our average height and the population of Earth.

- A map of our solar system, with Earth displaced toward the human figure to indicate our home world.
- A graphic of the Arecibo telescope itself, with its diameter.

The message was aimed at Messier 13 (M13), a dense globular star cluster some 25,000 light-years away. The choice was largely symbolic; M13 is a dense collection of stars, increasing the chances that someone might be there to receive it. Of course, since the message is traveling at the speed of light, it will take 25,000 years to arrive, and any potential reply would take another 25,000 years to get back. The Arecibo Message was not intended to start a conversation. It was a demonstration of our technological capability and a statement of our cosmic ambition.

The Golden Records: An Interstellar Time Capsule

While the Arecibo Message was a targeted broadcast, humanity has also sent physical artifacts into the cosmos. In 1977, NASA launched the twin Voyager 1 and 2 space probes on a grand tour of the outer solar system. Attached to the side of each spacecraft is a golden-plated copper phonograph record, an interstellar time capsule officially titled "The Sounds of Earth." The project was led by Carl Sagan, who assembled a small, brilliant committee to answer a monumental question: If you could tell the story of Earth to an unknown alien civilization, what would you say? The Golden Record was their answer.

The record's contents are a poignant and beautiful cross-section of humanity. They include:

- 115 images, encoded in analog form, showcasing scenes of nature, human anatomy, scientific diagrams, and everyday life across many cultures.
- A variety of natural sounds, such as wind, rain, thunder, birdsong, and whale calls.
- Musical selections from different cultures and eras, including Bach's Brandenburg Concerto, a Peruvian wedding song, Louis

Armstrong's "Melancholy Blues," and Chuck Berry's "Johnny B. Goode."

- Spoken greetings in 55 languages, ancient and modern, beginning with Akkadian (spoken in Sumer about 6,000 years ago) and ending with the modern Chinese dialect of Wu.

The Voyager probes have since completed their primary missions and are now hurtling through interstellar space. Voyager 1 officially crossed the heliopause—the boundary of our sun's influence—in 2012, becoming the first human-made object to enter the space between the stars. They will wander the Milky Way for billions of years. The chances of them being intercepted by another civilization are infinitesimally small. But if they are ever found, they will serve as a testament to the beings who made them—a species that looked up at the Great Silence and dared to send a greeting.

These early efforts—Ozma's listening, Arecibo's shout, and Voyager's whispered greeting—represent humanity's first forays into the cosmic conversation. They were born of a profound optimism that if we only knew how to look, and where to listen, we would find that the silence was not empty after all. The next step was to formalize the problem and attempt to calculate the odds, a task that would lead Frank Drake to create his most enduring legacy.

Part II: The Mathematical Probability

5

Chapter 5: The Drake Equation: An Introduction

How many intelligent, communicative civilizations are out there? Is our galaxy silent because life is rare, or are we simply not listening correctly? This is the central question of the SETI endeavor. In the early days of the search, it became clear that a structured way of thinking about the problem was needed—a method to break down this enormous question into smaller, more manageable pieces. The tool created for this purpose is perhaps the most famous and debated formula in astrobiology: the Drake Equation.

It is crucial to understand that the Drake Equation is not a mathematical proof or a precise calculation. It is a probabilistic argument, a simple way to organize our ignorance. Its purpose is not to provide a definitive number, but to provide a framework for conversation and scientific inquiry. It shines a light on what we know, and more importantly, what we don't. As Dr. Carl Sagan once praised, "It involves subjects ranging from stellar and planetary astronomy to organic chemistry, evolutionary biology, history, politics, and abnormal psychology. Much of the Cosmos is in the span of the Drake Equation."

The Man and the Meeting

The equation was conceived by Dr. Frank Drake in 1961. In the wake of his pioneering Project Ozma, Drake was tasked with organizing a small, high-level scientific meeting at the Green Bank observatory. The goal was to bring together the world's leading experts to discuss the burgeoning field of searching for extraterrestrial intelligence. The attendees were a "who's who" of science, including a young Carl Sagan, neuroscientist John C. Lilly (who was attempting to communicate with dolphins), and Melvin Calvin, who had just won the Nobel Prize for his work on photosynthesis.

To prepare an agenda for the meeting, Drake sought a way to structure the discussion. He began listing the key factors that would determine the number of civilizations we might be able to detect. Starting with the number of stars in the galaxy, he multiplied it by the fraction of those stars that have planets, then the fraction of those planets that are suitable for life, and so on. In about ten minutes, he had sketched out the equation that would bear his name. It was intended simply as an agenda, a way to ensure they covered all the relevant topics. It became his most enduring legacy.

The Equation Itself

The Drake Equation is typically written as follows:

$$N = R_* \times f_p \times n_e \times f_l \times f_i \times f_c \times L$$

Let's break down what each of these seven variables represents:

- **N:** This is the result we are trying to estimate. It represents the number of active, technologically advanced civilizations in the Milky Way galaxy whose electromagnetic signals are currently detectable.

- **R*** (The Rate of Star Formation): This is the average rate at which new stars are born in our galaxy. It sets the baseline, the total number of opportunities for solar systems to form.
- **fp** (The Fraction of Stars with Planets): Of all the stars being formed, what percentage of them host planetary systems? For most of the 20th century, this was a complete unknown.
- **ne** (The Number of "Earth-like" Planets): For each star that has planets, what is the average number of planets that are environmentally suitable for life to arise? This is often called the "Goldilocks" factor—planets that are not too hot, not too cold.
- **fl** (The Fraction of Planets Where Life Actually Arises): Of all the suitable planets, on what percentage of them does life actually begin? This variable represents the leap from non-living chemistry to living biology (abiogenesis).
- **fi** (The Fraction of Life-Bearing Planets Where Intelligence Arises): Of all the planets with life, on what percentage of them does life evolve to become intelligent? This accounts for the step from simple organisms, like bacteria, to complex creatures capable of abstract thought.
- **fc** (The Fraction of Intelligent Civilizations That Become Communicative): Of all the intelligent species, what percentage develop the technology and the desire to send detectable signals into space, such as radio waves?
- **L** (The Longevity of a Communicative Civilization): This is the average length of time that a communicative civilization survives and continues to send signals into space. This is often considered the most uncertain and, perhaps, the most important variable.

A Tale of Two Answers: Optimism vs. Pessimism

The power and frustration of the Drake Equation lie in its sensitivity to our assumptions. By plugging in different values for each variable, one can arrive at vastly different conclusions. In 1961, at the Green

Bank conference, Drake and his colleagues—including Sagan—plugged in their best "optimistic" estimates for each variable. Their values were roughly:

- R* = 1 star per year
- fp = 0.5 (half of all stars have planets)
- ne = 2 (two habitable planets per system)
- fl = 1 (every habitable planet develops life)
- fi = 0.01 (1% of life-bearing planets develop intelligence)
- fc = 0.01 (1% of intelligent civilizations become communicative)
- L = 10,000 years

Plugging these numbers in gives: N = 1 × 0.5 × 2 × 1 × 0.01 × 0.01 × 10,000 = 10.

A result of 10 detectable civilizations was encouraging. It suggested that a dedicated search might just succeed. But even this small change could drastically alter the outcome. If the longevity (L) was assumed to be, say, 100 million years, then N would become 100,000. If it was only 100 years, N would be 0.1, meaning we are likely alone in the galaxy at this moment.

In the decades since, our understanding of the first few variables has improved dramatically. But the last four—fl, fi, fc, and especially L—remain almost entirely in the realm of speculation. A "pessimistic" viewpoint might argue that the origin of life (fl) is incredibly rare, or that the average lifespan of a civilization (L) is short, as they tend to destroy themselves. A pessimistic estimate might look like this:

- R* = 1
- fp = 1 (we now know planets are everywhere)
- ne = 0.1 (truly Earth-like worlds are still rare)
- fl = 0.00000001 (abiogenesis is a near-miracle)
- fi = 0.001 (intelligence is very rare)

- fc = 0.01
- L = 500 years

This yields: N = 1 × 1 × 0.1 × 0.00000001 × 0.001 × 0.01 × 500 = 0.00000000005.

This result is a number far less than one, implying that we are almost certainly the only technological civilization in our galaxy, and perhaps even in the entire observable universe.

The Drake Equation does not give us the answer. Instead, it gives us a roadmap for the questions we need to ask. It tells us that to solve the Fermi Paradox, we must venture into astronomy, biology, geology, sociology, and futurology. In the following chapters, we will journey through the first few variables of this equation, replacing the speculation of 1961 with the hard-won scientific data of the 21st century. We will see that while some parts of the equation have come into sharp focus, others remain as mysterious as ever.

6

Chapter 6: R* & fp – Stars and Their Planets

The Drake Equation begins its grand calculation not with life, but with its necessary prerequisites: stars and the planets that orbit them. The first two variables, R* (the rate of star formation) and fp (the fraction of stars with planets), form the astronomical foundation of our search. In 1961, when Frank Drake first wrote his equation, these numbers were educated guesses, shrouded in uncertainty. Today, thanks to a golden age of astronomy, we have brought these first two terms into stunningly sharp focus. What we have learned has been nothing short of revolutionary, dramatically boosting the odds in life's favor.

R*: The Galactic Nursery

Look up at a clear, dark sky and you might see a faint, hazy patch in the sword of the Orion constellation. This is the Orion Nebula, a stellar nursery. It is a vast, turbulent cloud of gas and dust, more than twenty light-years across, where new stars are being born right now. Deep within these vibrant, glowing clouds, gravity is pulling matter together into dense, hot cores. When these cores become massive enough, the pressure at their centers ignites the fires of nuclear fusion, and a new star begins to shine. This process is happening all across our galaxy.

The variable R* represents the average rate of this process for the entire Milky Way. How many new stars does our galaxy produce each year? In 1961, Drake and his colleagues used an estimate of about ten stars per year, based on the total number of stars in the galaxy divided by its age. Modern astronomers, using data from telescopes that can peer into these dusty nurseries, have refined this number. The current consensus is that the Milky Way converts about 1.5 to 3 solar masses of gas into stars each year. Since many stars are smaller and less massive than our sun, this translates to an average of roughly three to five new stars igniting annually.

While this number seems small, it's a steady, relentless process of creation. Over the galaxy's 13-billion-year history, this stellar forge has produced the hundreds of billions of stars we see today.

But not all stars are created equal when it comes to hosting life. The type of star matters immensely.

- **O- and B-type Stars:** These are the giants—massive, incredibly hot, and brilliantly blue. They burn through their fuel at a furious rate and live for only a few million years before exploding in cataclysmic supernovae. This is likely far too short a time for complex life to evolve on any orbiting planets.
- **G-type Stars (like our Sun):** These are the yellow dwarfs. They are stable, long-lived (about 10 billion years), and provide a relatively consistent output of energy, making them ideal candidates for hosting life-bearing worlds.
- **M-type Stars (Red Dwarfs):** These are the most common type of star in the galaxy, making up about 75% of the stellar population. They are small, cool, and incredibly long-lived, with lifespans measured in the trillions of years. This would seem to give life an enormous amount of time to develop. However, they present their own challenges. They are prone to violent flares of radiation that could sterilize nearby planets, and any planet close enough to be warm would likely be tidally locked, with one side perpetually facing the star in unending daylight and the other frozen in eternal night.

For now, the key takeaway is that the raw material for solar systems is abundant. Every year, several new stars, many of them stable and long-lived, are added to the galactic inventory, each one a potential home for a new family of planets. But do they actually have them?

fp: A Revolution in Astronomy

For most of human history, we knew of only eight planets—the ones in our own solar system. We had no idea if our cosmic home was a standard model or a bizarre anomaly. When Frank Drake estimated fp, the fraction of stars with planets, as 0.5 (or 50%), he was making a bold but complete guess. There was simply no data.

The last three decades have changed everything. The story of how we went from knowing of eight planets to knowing of thousands is one of the great triumphs of modern science. The hunt for these "exoplanets" has relied on two primary, ingenious methods.

The first is the radial velocity method, sometimes called the "wobble method." As a planet orbits a star, its gravity doesn't just pull on the planet; the planet's gravity also pulls on the star, causing it to "wobble" slightly. By observing a star's light, astronomers can detect this tiny, periodic shift in its spectrum—a Doppler shift—as it moves slightly toward and away from us. In 1995, Swiss astronomers Michel Mayor and Didier Queloz announced the discovery of 51 Pegasi b, the first exoplanet found orbiting a sun-like star, using this very method. It was a watershed moment that opened the floodgates and later earned them the Nobel Prize in Physics.

The second, and even more prolific, method is the transit method. This technique doesn't look for a star's wobble, but for a minuscule dip in its brightness. If a planet's orbit happens to be aligned just right from our point of view, it will pass directly in front of its star, causing a brief, slight dimming—like a moth flying in front of a distant spotlight. NASA's Kepler Space Telescope, launched in 2009, was a revolutionary mission designed specifically for this purpose. For four years, it stared relentlessly at a single patch of sky containing over 150,000 stars, measuring their brightness with unprecedented precision.

The results from Kepler, and its successor TESS, have been staggering. They have not only confirmed thousands of exoplanets but have fundamentally rewritten our understanding of fp. Drake's optimistic guess of 0.5 turned out to be far too conservative. The current data strongly indicates that the value of fp is approximately 1.

This is a profound discovery. It means that, on average, there is at least one planet for every star in the galaxy. Planets are not the exception; they are the rule. The Milky Way is home to not millions, but hundreds of billions, and perhaps even trillions, of worlds. They come in a bewildering variety, from gas giants larger than Jupiter orbiting scorching-hot, close to their stars ("Hot Jupiters") to rocky worlds several times the mass of our own ("Super-Earths").

The first two terms of the Drake Equation are now solved with a surprising degree of certainty. We know that the galaxy is a tireless manufacturer of stars (R*) and that these stars are almost all furnished with their own planetary systems (fp). The raw real estate for life is more abundant than even the optimists of 1961 dared to hope.

This knowledge transforms the Fermi Paradox. It eliminates the easy answer that planets might be rare. The galaxy is teeming with worlds. This forces the burden of proof further down the equation. It's not enough to have planets; you need the right kind of planets. Our next step, then, is to journey to the "Goldilocks Zone" and ask: of these trillions of worlds, how many are actually capable of supporting life?

7

Chapter 7: ne - Worlds Fit for Life

The discovery that planets are as common as stars is a seismic shift in our cosmic perspective. It has transformed the galaxy from a sparsely decorated void into a treasure trove of trillions of worlds. The real estate for life is undeniably vast. But as any realtor will tell you, location is everything. Most of these worlds are wholly unsuitable for life as we know it—scorched gas giants, barren ice balls, and airless rocks. The variable ne in the Drake Equation forces us to move from a simple planet census to a more discerning search. It asks: For each star that has planets, what is the average number of worlds that are environmentally suitable for life to arise?

This is the search for "Goldilocks" planets—worlds that are not too hot, not too cold, but just right. Determining the criteria for this perfect cosmic address is one of the most active and exciting fields in modern astrobiology.

The "Goldilocks Zone": A Place in the Sun

The most fundamental requirement for life as we know it is liquid water. Water is the universal solvent, the medium in which the chemical reactions of biology take place. It facilitates the transport of nutrients and waste and plays a key role in regulating temperature. While scientists speculate about exotic life based on other solvents like liquid methane or ammonia, our search must begin with what we know works. And what works is water.

For a planet to host liquid water on its surface, it must orbit its star at just the right distance. This orbital sweet spot is known as the Circumstellar Habitable Zone (CHZ), or more popularly, the "Goldilocks Zone."

- If a planet is too close to its star, the intense energy will cause its oceans to boil away. Any water vapor in the atmosphere would trigger a runaway greenhouse effect, creating a crushing, furnace-like world. We need only look next door to Venus, with its surface temperature hot enough to melt lead, to see a stark example of this fate.
- If a planet is too far from its star, it will be too cold. Its oceans will freeze solid, locking away the essential solvent for life. Mars, a frigid desert world on the outer edge of our own system's habitable zone, shows us this cold reality.

Earth orbits comfortably within the sun's habitable zone. This simple fact of cosmic geography is the primary reason our world is a vibrant, water-rich biosphere and not a sterile inferno or a desolate ice ball.

Beyond Goldilocks: It's More Than Just Temperature

Location in the habitable zone is a crucial first step, but it is not a guarantee of habitability. A planet can be in the perfect spot and still be utterly hostile to life. Several other factors are critical.

First among them is atmospheric pressure. Liquid water can only be stable under a sufficient atmosphere. On a planet with very little air, like Mars today, the pressure is so low that liquid water on the surface would instantly boil, even at cold temperatures. To have stable oceans, lakes, and rivers, a planet needs a substantial atmosphere to press down on the water and keep it in its liquid state. This, in turn, implies that the planet must have enough mass, and therefore gravity, to hold on to its atmosphere over billions of years. A world much smaller than Earth, even if it's in the habitable zone, will likely have its atmosphere slowly sputtered away by the solar wind.

Second, we are discovering that the surface is not the only place to look for water. This realization has dramatically expanded our search, leading us to one of the most exciting prospects in our own solar system: the icy moons of Jupiter and Saturn. Moons like Jupiter's Europa and Saturn's Enceladus orbit far outside the sun's habitable zone, and their surfaces are covered in a thick shell of solid ice, with temperatures hundreds of degrees below zero.

Yet, evidence strongly suggests that beneath these frozen crusts lie vast, global oceans of liquid water. What keeps them from freezing? A process called tidal heating. As these moons orbit their colossal parent planets, the immense gravitational pull of Jupiter and Saturn constantly stretches and flexes their interiors. This friction generates enormous amounts of heat, enough to maintain a liquid ocean miles beneath the ice. Enceladus has even been observed actively spewing plumes of water vapor and ice crystals from its south pole—a direct sign of the liquid ocean below. These subsurface oceans are dark and under immense pressure, but they are protected from surface radiation and may contain hydrothermal vents on their seafloors, just like those on Earth where life is thought to have possibly begun. This discovery means that ne might not just include surface worlds, but potentially trillions of icy moons in the outer reaches of countless solar systems.

The Unseen Shields: A Planet's Defenses

A habitable world doesn't just need the right ingredients; it needs to protect them. Life is fragile. A planet capable of sustaining it must have defenses against the relentless dangers of space.

The most important of these is a global magnetic field, or magnetosphere. Generated by the motion of a molten iron core, a planet's magnetic field acts as an invisible force field, deflecting the solar wind—a constant stream of charged particles flowing from its star. Without a magnetosphere, this stellar wind would, over eons, strip a planet's atmosphere away, blow away its oceans, and sterilize its surface with radiation. Mars is a tragic example. It once had a thick atmosphere and liquid water, but after its small core cooled and its magnetic field died, the solar wind scoured it clean, leaving the barren desert we see today.

The atmosphere itself is the second critical shield. It burns up most small meteors, protecting the surface from constant bombardment. It also contains layers, like our ozone layer, that absorb the most harmful forms of ultraviolet radiation from the sun, which would otherwise damage DNA and make life on the surface impossible.

Putting it all together, our list of requirements for a truly "Earth-like" planet has grown. We need a rocky world with enough mass to hold a substantial atmosphere, orbiting in the habitable zone of a stable star, and possessing a molten core to generate a protective magnetic field. Despite these stringent requirements, the sheer number of planets in our galaxy means that the final tally for ne is still breathtakingly large. Based on the data from the Kepler mission, astronomers estimate that there could be as many as 10 to 40 billion potentially habitable, Earth-sized planets in the Milky Way alone. Even if only a fraction of these are truly habitable, the galaxy is still left with an enormous inventory of potential cradles for life.

The astronomical factors of the Drake Equation have given us an unequivocally optimistic result. The galaxy is rich with stable stars, overflowing with planets, and home to billions of worlds that could, in theory, host liquid water. The silence of the cosmos, therefore, cannot be easily blamed on a lack of suitable real estate. The problem must lie further down the equation, in the far more uncertain realms of biology and sociology. Having found an abundance of habitable stages, our next question is the most fundamental of all: how often does the play of life actually begin?

8

Chapter 8: fl, fi, & fc – From Slime to Signals

We have now crossed a critical threshold in our journey through the Drake Equation. The first three variables—R*, fp, and ne—belong to the realm of astronomy. They deal with stars, orbits, and planets, and while they involve immense numbers, they are governed by the relatively predictable laws of physics. We have moved from speculation to statistical certainty and found that the galaxy is rich with potentially habitable worlds.

But now we must leave the comforting certainty of physics and plunge into the murky, speculative waters of biology and sociology. The next three variables—fl (the fraction of habitable planets where life arises), fi (the fraction of those where intelligence evolves), and fc (the fraction of intelligent species that become communicative)—are almost complete unknowns. They are the heart of the uncertainty in the Drake Equation and the place where optimists and pessimists diverge most dramatically. These variables represent the three great evolutionary leaps that must occur to produce a civilization we can detect.

fl: The Genesis Factor

The variable fl confronts us with the mystery of abiogenesis. Of all the billions of "Goldilocks" planets in our galaxy, on what fraction of them does life actually ignite? Is the jump from non-living chemistry to a living, reproducing cell a common, almost inevitable process given the right ingredients, or is it a fluke event of such staggering improbability that it has happened only once? This is perhaps the most fiercely debated variable in the equation.

The optimistic view holds that fl is high, perhaps close to 1. Proponents of this view argue that life is not a miracle, but a natural outcome of cosmic chemistry. The building blocks of life—amino acids, nucleotides—are found in meteorites and are readily synthesized in laboratory experiments that simulate early Earth conditions. Given a suitable planet with liquid water and a source of energy, the laws of chemistry and thermodynamics will naturally drive matter toward greater complexity, eventually leading to self-replicating systems. The fact that life appeared on Earth very early in its history, almost as soon as the planet was cool enough to support it, is often cited as evidence. Life, in this view, is written into the fabric of the cosmos; the universe is hard-wired for biology. Those who support this hypothesis often argue that if the one example we have immediately formed life, than we must logically conclude that Earth did not constitute the exception, but simply followed the rule.

The pessimistic view, often associated with the Rare Earth Hypothesis, argues that fl is vanishingly small. While the building blocks may be common, assembling them into the first functioning cell is a hurdle of unimaginable difficulty. A living cell is a marvel of microscopic machinery, with thousands of proteins working in concert, all governed by a complex genetic code. The odds of the right molecules spontaneously assembling in the right order to form the first self-replicating entity, pessimists argue, are astronomical. They contend that Earth's genesis event might be a singular piece of luck in a mostly sterile galaxy. The logic behind this opinion is that even if I have a box containing metal ore and every other component necessary to build a watch, it's never going to randomly refine and assemble itself into a Rolex.

We have only one data point: Earth. The fact that life exists here tells us that fl is not zero. But that is all it tells us. Until we discover a second, independent origin of life—a "second genesis," perhaps on Mars or Europa—this variable will remain the subject of profound speculation.

fi: The Intelligence Factor

Let's assume life does arise. The next great hurdle is the evolution of intelligence. Of all the planets teeming with microbial slime, jungles, and oceans, on what fraction does a species evolve with the cognitive ability for abstract thought, language, and technology?

The optimistic case for fi often points to the principle of convergent evolution. This is the phenomenon where different, unrelated species independently evolve similar traits to adapt to similar environmental pressures. The eye, for instance, has evolved independently at least 40 different times in various branches of the animal kingdom. Flight evolved separately in insects, pterosaurs, birds, and bats. The argument is that intelligence is such a powerful evolutionary advantage—allowing for complex problem-solving, tool use, and social cooperation—that it is an inevitable outcome of evolution, given enough time and competition. If life is common, then intelligent life, sooner or later, is also common.

The pessimistic case is sobering. On the only planet we know, life has existed for nearly 4 billion years. In that entire time, out of the billions of species that have ever lived, only one—*Homo sapiens*—has developed high technology. If intelligence is such a clear advantage, why do we not share our planet with intelligent dolphins, tool-using crows, or philosophical octopuses? The dinosaurs ruled the Earth for 150 million years but never built a radio telescope. This suggests that the path to high intelligence is not a broad highway but a treacherous, narrow path riddled with evolutionary dead ends. Perhaps it requires a unique combination of factors: a large brain, grasping appendages (like hands), the development of complex language, and the harnessing of fire. The fact that it happened only once here on Earth strongly suggests that fi may be an extremely small number.

fc: The Communication Factor

Let us now assume that an intelligent species has evolved. The final biological leap is fc: what fraction of these species develop technology that is detectable across interstellar space? It is one thing to be intelligent; it is another to build a radio transmitter. To become a "communicative" species in the SETI sense, a civilization must have an interest in science and exploration. It must understand the laws of physics and electromagnetism. It must have the material resources and industrial capacity to build large transmitters.

One could argue that curiosity is a natural trait of intelligence, and that any intelligent species would eventually look to the stars and try to understand its place in the cosmos, leading inevitably to physics and radio astronomy. In this view, fc would be relatively high. However, it's also possible to imagine scenarios where this doesn't happen. What if the dominant intelligent species on a planet were aquatic, like dolphins or octopuses? Living in a dense, watery medium, they might never look to the sky or discover fire, the prerequisite for metallurgy and eventually electronics. They could be brilliant poets and philosophers but would remain forever silent to the wider cosmos. Or what if a civilization's philosophical or religious systems actively discouraged scientific inquiry or looking beyond their own world? It's entirely possible for a species to be intelligent but not technological.

These three variables—fl, fi, and fc—represent the biological and sociological filters that stand between a habitable planet and a detectable signal. Unlike the astronomical terms, they cannot be easily solved with better telescopes. They depend on the contingent, unpredictable, and perhaps unique evolutionary path taken on each world. Even if the galaxy is filled with habitable planets (high ne), if any one of these three "f" factors is infinitesimally small, the galaxy will fall silent. If abiogenesis is a near-miracle (low fl), or if intelligence is a freak accident (low fi), then the silence is explained. But there is one final variable, a term that haunts all discussions of the Fermi Paradox. It acts as a final multiplier, capable of rendering all the previous optimistic estimates null and void. This is the great question of longevity.

9

Chapter 9: L – The Great Timespan

We arrive now at the final and most haunting variable of the Drake Equation: L, the longevity of a communicative civilization. This single letter represents the average length of time that a technological society survives and continues to send detectable signals into space. Unlike the other variables, which deal with origins and fractions, L deals with endings. It is the term that forces us to confront our own fragility and the immense, soul-crushing scale of cosmic time.

You can be an extreme optimist about every other factor in the equation. You can believe that every habitable planet sprouts life (fl=1), that intelligence is a cosmic imperative (fi=1), and that all intelligent beings eventually build radio telescopes (fc=1). But if the value of L is short, the Great Silence is the inevitable result. L is the great bottleneck, the final filter that can transform a bustling, vibrant galaxy into a cosmic graveyard of fleeting, long-dead civilizations. Pondering L for too long often forces us to confront one of the most frightening questions humanity has to face: How long do we think we'll survive as a species?

The Analogy of the Cosmic Ocean

To understand the profound power of L, imagine the history of the Milky Way as a vast ocean stretching for 13 billion miles, where one mile equals one year. Our entire recorded human history, from ancient Sumer to the present day, would occupy less than the last three inches of this ocean. The time we have been a "communicative" civilization, capable of sending radio signals, is less than the thickness of a single human hair.

Now, imagine that somewhere in this vast ocean, other ships (civilizations) are lighting a flare (becoming communicative). The flare only stays lit for the length of time they survive and remain technological—their value of L. If L is very long, say 100 million years (100 million miles in our ocean), then the ocean would be filled with the overlapping lights of many flares, and the chance of seeing one would be high. But if L is short, say 500 years (a mere 500 miles), then each flare is a tiny, instantaneous flash in a vast, dark expanse. Even if millions of ships have lit flares over the ocean's history, the chance of any two flares being lit at the same time and in a position to see each other becomes vanishingly small. This is the crux of the problem. For us to detect another civilization, our brief technological window must overlap with theirs across billions of years of cosmic history.

The Dangers of a Short L: Existential Threats

Why would L be short? The answer lies in the concept of existential risk—a threat that could cause the extinction of intelligent life or permanently and drastically curtail its potential. Unfortunately, the very technology that makes a civilization communicative (high energy physics, rocketry, global communication) also tends to give it the power to destroy itself.

This is the essence of the Self-Destruction Hypothesis, which posits that the lifespan of technological civilizations is inherently short. Consider humanity's own trajectory. We developed the capability for self-annihilation via nuclear weapons in 1945, decades before we took our first steps on another celestial body. We have been a "communicative" species for only about a century, yet in that time we have faced, and continue to face, a host of self-inflicted existential threats:

- **Nuclear War:** A full-scale thermonuclear exchange could not only kill billions but also trigger a "nuclear winter" that would collapse the global ecosystem, potentially leading to extinction.
- **Climate Change:** Unchecked global warming could lead to runaway feedback loops, rendering large parts of the planet uninhabitable, causing societal collapse and the loss of our technological infrastructure.
- **Engineered Pandemics:** As biotechnology advances, the potential for a genetically engineered pathogen—either accidentally released or deliberately used as a weapon—with high lethality and transmissibility poses a catastrophic threat.
- **Uncontrolled Artificial Intelligence:** A superintelligent AI, if not properly aligned with human values, could pursue its programmed goals in ways that are destructive to its creators, seeing us as an obstacle or a resource to be consumed.

Beyond these self-inflicted risks, civilizations also face external threats:

- **Asteroid or Comet Impacts:** The event that wiped out the dinosaurs was not a one-time occurrence. Large impacts are a recurring, if infrequent, feature of planetary history. A civilization must be technologically advanced enough to detect and deflect such a threat to ensure its long-term survival.
- **Supervolcanoes and Other Geologic Catastrophes:** A massive eruption could blanket the planet in ash, blocking out sunlight for years and causing a global famine and societal collapse.
- **Nearby Supernovae:** If a nearby massive star were to explode, it would bathe our planet in lethal levels of radiation, potentially stripping away our atmosphere and sterilizing the surface.

The sobering reality is that a civilization must successfully navigate a minefield of these threats not just for a century, or a millennium, but for the millions of years required to become a truly long-lived, stable presence in the galaxy. The path to a large L is perilous. The slightest misstep can reset the clock to zero.

The Hope for a Long L

Is a short L inevitable? Not necessarily. An optimist might argue that the very intelligence that creates these threats also provides the means to overcome them. A civilization that survives its technological adolescence might achieve a state of virtual immortality. Having overcome the risks of war, climate change, and pandemics, such a society might develop into a "post-human" or Type II civilization (on the Kardashev scale), capable of harnessing the full energy output of its parent star. They would be immune to most natural disasters, capable of deflecting asteroids, controlling their planet's climate, and may have even uploaded their consciousness into more durable, non-biological forms. Such a civilization could have an L measured in millions or even billions of years, limited only by the lifespan of its star.

If even a handful of civilizations achieve this state of stability, they would become cosmic fixtures, akin to lighthouses shining across the ocean of time. Their signals would be broadcasting for so long that we would be almost guaranteed to detect one, even with our own brief listening window.

The value of L is the great unknown that governs the entire equation. If L is typically short—a few hundred or a few thousand years—then the Great Silence is the sad, logical consequence of civilizations blinking in and out of existence, forever alone in the vast darkness. But if even one civilization in a million manages to survive its technological infancy and achieve a long L, then the paradox returns in full force. Where are the immortals? Where are the lighthouses?

With this final, troubling variable, our analysis of the Drake Equation is complete. We have found that the galaxy is filled with stages (planets), but we do not know how often the actors (life, intelligence) take the stage, or for how long their play (civilization) runs. The equation has not given us the answer, but it has brilliantly framed the question. It shows us that the Great Silence is the product of either a scarcity of beginnings or a plethora of endings. We now turn from calculating the odds to exploring the specific hypotheses that attempt to explain which of these is the truth.

Part III: The Proposed Solutions

Having quantified the mystery with the Drake Equation, we now arrive at the heart of our investigation. The astronomical variables suggest a galaxy rich with habitable real estate, yet the silence persists. This implies that the solution to the Fermi Paradox lies in the more speculative, and far more interesting, realms of biology, sociology, and even metaphysics.

In this section, we will embark on a comprehensive tour of the most compelling, unsettling, and imaginative explanations ever proposed for the Great Silence. Each hypothesis acts as a unique lens through which to view the paradox, offering a different reason for the universe's apparent emptiness. We will weigh the lonely possibility that our world is an extraordinary accident, confront the terrifying idea that a "Great Filter" dooms most life, and entertain the paranoid logic of a cosmos where silence is the only strategy for survival. We will also explore more exotic ideas: that we are being deliberately isolated in a galactic "Zoo," that our entire reality is a computer program, or that life exists in dimensions our senses cannot perceive.

These are not just scientific hypotheses; they are narratives that force us to confront our deepest assumptions about life, intelligence, and humanity's place in the cosmos. Let us now step into the gallery of solutions and examine the evidence for each.

10

Chapter 10: The Rare Earth Hypothesis

The principle of mediocrity is a foundational concept in cosmology. It suggests that our planet, our sun, and our galaxy are typical, and that there is nothing special about our place in the universe. It is this principle that underpins the optimistic belief that if life happened here, it must have happened in countless other places. The Rare Earth Hypothesis launches a direct and powerful assault on this idea. Maybe we are a true cosmic anomaly.

First proposed in the 2000 book *Rare Earth: Why Complex Life Is Uncommon in the Universe* by paleontologist Peter Ward and astronomer Donald Brownlee, this hypothesis argues that while simple microbial life might indeed be common, the specific set of circumstances that allowed for the evolution of complex, multicellular animal life on Earth are so exceptionally rare that our planet might be a unique, or near-unique, anomaly in the galaxy.

The hypothesis does not argue that Earth itself is the miracle. Instead, it argues that Earth is the result of a long, improbable chain of fortunate accidents. A single broken link in this chain would have left our world as just another sterile rock or, at best, a planet of slime. Let's examine the key links in this chain of cosmic luck.

1. **The Right Location in the Right Galaxy**

Our journey begins on the largest scale: the galaxy. Not all parts of the Milky Way are equally hospitable. We reside in what is called the Galactic Habitable Zone, a narrow, stable ring of stars part-way out from the galactic center.

- **Too close to the center:** The galactic core is a violent, chaotic region. The density of stars is immense, increasing the likelihood of gravitational disruptions to planetary orbits. More dangerously, the supermassive black hole at the center, Sagittarius A*, and the frequent nearby supernovae would bathe any inner-galaxy planets in sterilizing levels of high-energy radiation.
- **Too far from the center:** The outer regions of the galaxy are much calmer but are also "metal-poor." In astronomy, "metals" are any elements heavier than hydrogen and helium. These heavier elements—carbon, oxygen, iron, silicon—are the essential building blocks of rocky planets and of life itself. They are forged in the hearts of stars and scattered through space in supernova explosions. The outer galaxy, having had fewer generations of stars, is deficient in these materials, making the formation of Earth-like planets less likely.

Our solar system is perfectly situated in this calm, metal-rich suburban ring, a location that is both safe and well-stocked with the necessary ingredients for life.

2. **The Right Kind of Star**

As we discussed in Chapter 6, our G-type sun is a model of stability. It has a lifespan of about 10 billion years, providing a long, stable period for complex life to evolve. It doesn't erupt with the violent, sterilizing flares common to the more numerous red dwarf stars. Its energy output is relatively consistent, preventing the wild temperature swings that would doom nascent life.

3. **The Right Planetary Arrangement**

   Our solar system's architecture is crucial. The presence of Jupiter, a colossal gas giant in the outer solar system, has been indispensable for our survival. Jupiter acts as a "cosmic bouncer," a gravitational shield that has ejected countless comets and asteroids from the inner solar system or absorbed their impacts. Without Jupiter's protection, Earth would have been subjected to a rate of catastrophic impacts ten thousand times greater, likely preventing complex life from ever gaining a foothold.

4. **The Guardian Moon**

   One of the most unusual features of our planet is its moon. Compared to its parent planet, Earth's Moon is exceptionally large. Most moons are tiny, captured asteroids or debris. Our Moon, according to the prevailing "Giant-Impact Hypothesis," was formed when a Mars-sized protoplanet named Theia collided with the early Earth. This cataclysmic event sprayed a massive ring of debris into orbit, which eventually coalesced into our Moon.

   This violent birth had two crucial consequences. First, the impact is thought to have blasted away Earth's primordial, hydrogen-helium atmosphere, making way for the secondary atmosphere of volcanic outgassing that would eventually become breathable. Second, and most importantly, our large Moon acts as a powerful gravitational anchor. It stabilizes Earth's axial tilt, the angle of its rotation. Without the Moon, gravitational perturbations from Jupiter and the Sun would cause Earth's tilt to swing wildly over millions of years, leading to extreme and unpredictable climate changes that would be devastating to complex life. Our stable seasons and relatively mild climate are a direct gift from our guardian Moon.

5. **The Engine of Plate Tectonics**

   Perhaps the most subtle but critical factor is Earth's internal geological engine: plate tectonics. Earth's crust is broken into several large plates that are in constant, slow motion. This process is vital for life in several ways:

- It acts as a global thermostat. Through volcanic activity, it releases carbon dioxide into the atmosphere, creating a greenhouse effect that keeps the planet warm. But it also removes CO2 through the weathering of rocks, which are then subducted back into the mantle. This carbon cycle has regulated Earth's climate for billions of years, preventing it from freezing over or overheating.
- It creates continents and promotes biodiversity. The constant shifting of landmasses creates diverse habitats and drives evolutionary innovation.
- It drives the magnetic field. The convection within the mantle, which drives plate tectonics, is linked to the convection in the liquid outer core that generates our protective magnetosphere.

It is not known how common plate tectonics are on other rocky worlds. Venus has none. Mars appears to have none. It may be another rare feature that distinguishes our living world from the dead ones.

The Rare Earth Hypothesis argues that when you multiply the probabilities of all these factors—and many more—occurring for a single planet, the result is a number astonishingly close to zero. We are not mediocre; we are a perfect storm of improbabilities. We are a cosmic jackpot. If this hypothesis is correct, the Great Silence is not a paradox at all. It is the natural sound of a universe where simple microbial slime may be common, but worlds like ours—teeming with the beautiful, complex, and intelligent tapestry of animal life—are extraordinarily rare and precious jewels, separated by unimaginable voids of sterile space.

11

Chapter 11: The Great Filter Hypothesis

If the Rare Earth Hypothesis is a story about our planet's unique good fortune, the Great Filter Hypothesis is its darker, more universal counterpart. It is less concerned with the specific conditions of a single planet and more concerned with the overarching evolutionary and technological path that any form of life must travel to reach a state of interstellar communication. The hypothesis posits that this path is fraught with immense, almost insurmountable barriers, and that somewhere along this path lies a "Great Filter."

The term was coined in a 1998 essay by economist Robin D. Hanson. He argued that the eerie silence of the cosmos implies that at least one of the steps required to get from a dead planet to a galaxy-spanning civilization is extraordinarily improbable. This step is the Great Filter. It is the hurdle that the vast, vast majority of life-bearing planets fail to overcome.

The chilling power of this hypothesis lies in its implications for humanity. The crucial question is not just "What is the filter?" but "Where is the filter in relation to us?" Are we one of the lucky few who have already passed it, or is the greatest challenge, the one that dooms almost everyone, still lying in our future? Let's walk the path of a potential civilization and examine the primary candidates for the Great Filter.

The Path to Expansion

We can lay out a simplified, nine-step evolutionary path that must be completed for a species to become a detectable, space-faring civilization:

1. The Right Star System: A habitable planet in a stable orbit.
2. Reproductive Molecules: The emergence of self-replicating molecules (e.g., RNA).
3. Prokaryotic Life: The development of simple, single-celled life.
4. Eukaryotic Life: The evolution of complex cells with specialized organelles.
5. Sexual Reproduction: The evolution of a method for combining genetic information, vastly accelerating evolutionary adaptation.
6. Multicellular Life: The development of complex organisms with specialized tissues and organs.
7. Tool-Using Intelligence: The evolution of animals with large brains and the capacity for complex tool use (like early humans).
8. Technological Civilization: A civilization that develops science, industry, and global communication (where we are now).
9. Interstellar Colonization: The expansion into space, making the civilization immune to planetary-scale disasters.

According to the logic of the Great Filter, at least one of these steps must be incredibly improbable. The fact that our universe appears dead suggests that somewhere on this list is a wall that almost no one ever climbs.

Possibility 1: The Filter is Behind Us

This is the optimistic scenario. It suggests that one of the early steps on the path was the incredibly difficult one, and we, through some cosmic lottery, have already passed it. If this is true, it means humanity is exceptionally rare and precious, and we may have a clear path to a vast and empty galaxy.

There are two primary candidates for a "past" filter:

- **Candidate A: Abiogenesis (Step 2 or 3).** The Great Filter could be the very origin of life itself. As discussed in Chapter 3, we still do not know how non-living chemistry sparked into a living, self-replicating system. It's possible that the conditions must be so perfect, the spontaneous chemical reactions so unlikely, that it has happened only a handful of times in the galaxy's history. In this case, our planet is one of the winning tickets. We might find sterile, habitable worlds everywhere, but worlds with even the simplest bacterium are cosmic miracles.
- **Candidate B: The Leap to Complex Cells (Step 4).** This is a subtler but equally compelling argument. For the first two billion years of life on Earth, the planet was inhabited exclusively by simple prokaryotic cells, like bacteria. Then, in a single, bizarre evolutionary event, one cell is thought to have engulfed another in a symbiotic relationship, giving rise to the first complex eukaryotic cell. This new type of cell contained mitochondria (the powerhouses of the cell, descendants of the engulfed bacterium) and a nucleus to house its DNA. Every single plant, animal, and fungus on Earth is a eukaryote. This event happened only once in 4 billion years. It's possible that this leap from simple to complex life is the true bottleneck. The universe might be teeming with "slime worlds" of bacteria, but almost none of them ever make the leap to complexity, and thus can never evolve intelligence.

If the filter is behind us, the Great Silence is a sign of our own rarity. It means the cosmos is a vast, empty inheritance waiting for us to claim it. This thought is both lonely and exhilarating. It may be the case that we are among the few civilizations proven capable of passing the universe's arduous tests.

Possibility 2: The Filter is Ahead of Us

This is the terrifying scenario. It suggests that the early steps are relatively easy, and that life and even intelligence may be common throughout the universe. But if this is the case, why is the galaxy silent? Because some future step, some predictable challenge, consistently destroys civilizations before they can master interstellar travel. If this is true, then the silence is a warning. It is the sound of other civilizations reaching the same stage we are at now and failing to survive. Perhaps humanity's real challenges have yet to come.

What could this future filter be? The candidates are the very existential risks we discussed in the previous chapter on Longevity (L).

- **The Inevitability of Self-Destruction (Step 8 to 9):** This is the most common argument for a future filter. It posits that the development of technology outpaces the development of social and ethical maturity. A civilization discovers the power of the atom before it has abolished war. It develops world-altering industrial processes before it understands their long-term consequences on the climate. It creates powerful artificial intelligence before it can guarantee its control. The Great Filter, in this view, is the near-certainty that a civilization will destroy itself with the very power it sought to master. The silence we hear is the silence of countless nuclear winters, runaway greenhouse effects, and rogue AIs. We are currently in the midst of this filter.
- **The Great Exhaustion:** Another possibility is that civilizations don't destroy themselves in a single cataclysm, but simply exhaust their resources and collapse. A technological society has immense energy and resource needs. Perhaps most planets simply don't have enough easily accessible fossil fuels or rare earth metals to sustain an industrial phase long enough to make the leap to fully renewable energy and interstellar capability. The civilization flourishes for a few centuries, burns through its planetary inheritance, and then collapses back into a pre-industrial state, unable to ever make the leap again.

If the Great Filter is ahead of us, then the silence of the cosmos is a deeply ominous sign. It suggests that we are likely doomed. Every civilization before us has reached this point and perished, and the statistical odds are overwhelmingly against us being the first to succeed.

Hanson's hypothesis forces a stark choice. Which is more likely: that we are the first and only ones to have made it this far, or that we are simply the next in a long line of civilizations destined for failure? When we look at the stars and hear only silence, we may be looking at our own inevitable future.

12

Chapter 12: The Dark Forest Theory

The solutions to the Fermi Paradox we have explored thus far—Rare Earth, the Great Filter—have one thing in common: they assume the silence exists because the galaxy is empty. They are explanations for a lack of life. But what if that assumption is wrong? What if the galaxy is teeming with intelligent civilizations, and the silence exists not because of absence, but because of fear?

This is the premise of the Dark Forest Theory, one of the most unsettling and compelling explanations for the Great Silence. The theory was not born from a scientific paper, but from the pages of science fiction, specifically from the 2008 novel *The Dark Forest* by Chinese author Liu Cixin. Yet, its internal logic is so ruthlessly game-theoretical that it has been embraced by many as a terrifyingly plausible model for cosmic behavior. The theory asks us to re-imagine the universe not as an empty space waiting to be explored, but as a dark forest at midnight.

The Analogy of the Forest

Imagine a dark forest, filled with countless armed hunters. Each hunter must tiptoe silently through the trees, pushing aside branches with the utmost care, trying not to make a sound. Even their breathing is cautious. Why? Because the forest is full of other hunters, just as silent and stealthy as they are. If a hunter discovers another living being—whether it's another hunter, a helpless infant, or a benevolent angel—there is only one thing they can do: open fire and eliminate them.

In this forest, any creature that reveals its own existence is swiftly wiped out. It is a world of kill-or-be-killed, where survival depends on remaining invisible. This, Liu Cixin argues, is the true state of our universe. The silence we perceive is not the silence of a graveyard; it is the deliberate, watchful silence of hunters holding their breath. Any fisherman will tell you that most game fish like cover where they can hide. Any hunter will tell you that most wild game bed down where they feel hidden and safe. Loudly announcing one's presence and actively seeking out the unknown are clearly not the most intelligent of survival tactics. If this pattern of self-preservation through stealthiness has emerged as a leading practice on Earth, why should life throughout the cosmos not follow a similar strategy?

The Axioms of Cosmic Sociology

This terrifying conclusion rests on two simple, foundational axioms about how any civilization would behave:

1. **Survival is the primary need of a civilization.** This is the most basic instinct of any life form. A civilization will do whatever is necessary to ensure its own continued existence.
2. **A civilization will continuously grow and expand, but the total matter in the universe remains constant.** Any civilization with technology will inevitably consume more resources and seek more living space. This puts all civilizations in a zero-sum game, competing for a finite pool of resources.

From these two axioms, two more crucial concepts emerge: chains of suspicion and the technological explosion.

Chains of Suspicion: Imagine two civilizations, A and B, that become aware of each other across the vastness of interstellar space. Can they communicate and live in peace? The Dark Forest theory says no. Because of the immense distances, communication is subject to crippling time lags. If A sends a message of "hello" to B, it might take thousands of years to arrive. By the time B receives it, has A's benevolent culture changed? Has it become hostile? B cannot be sure.

Furthermore, A cannot trust that B is truly peaceful. B might be feigning benevolence while preparing a first strike. A knows this, and B knows that A knows this. This leads to a recursive, unending "chain of suspicion." You can't be sure of another civilization's true intentions, and you can't be sure they will remain benevolent across the vast timescales of interstellar communication. Even if both civilizations are genuinely peaceful, neither can afford to believe the other is. The safest assumption is always that the other party is a threat.

The Technological Explosion: Compounding this problem is the unpredictable nature of scientific progress. A civilization that is technologically primitive now might, in a few hundred or a few thousand years (a cosmic eye-blink), undergo a "technological explosion" and surpass you. A species that seems harmless and backward today could become an existential threat tomorrow.

Consider the perspective of an advanced, million-year-old civilization that detects our own radio signals. To them, we are like a child playing with matches in a forest. We are currently no threat. But can they be sure that in a thousand years, we won't develop some new form of physics that could destroy them from across the galaxy? No. They cannot know our future technological trajectory. The only safe and logical course of action is to extinguish the match before it has a chance to start a wildfire.

The Terrifying Conclusion

When you combine the axioms of survival and expansion with the unavoidable realities of chains of suspicion and technological explosions, the conclusion is chillingly simple. The "survival of the fittest" model exhibited on Earth seems to fall in line with Dark Forest theory. Why wouldn't the cosmos likewise follow that logic? For any intelligent civilization, the moment it detects another civilization, it is faced with a choice:

1. Attempt to communicate. This is a massive risk. You reveal your own position and nature, and you can never be certain of the other's intentions or future capabilities.
2. Remain silent and hope they are friendly. This is also a massive risk. If they are hostile, you have done nothing to protect yourself.
3. Destroy them preemptively. This is the only logical, low-risk solution. By eliminating the other civilization before it can become a threat, you guarantee your own survival. The horror surrounding this line of thinking is that if we can arrive at such a conclusion - than so can they.

Therefore, the optimum strategy for every civilization in the cosmos is to remain hidden and silent while simultaneously listening for others. If you hear a signal, you immediately know the location of a naive and foolish civilization. You then have a cosmic obligation to your own survival to destroy them.

In this context, SETI is transformed from a hopeful, optimistic search into an act of spectacular, suicidal recklessness. Every message we broadcast, from the Arecibo Message to our television and radio leakage, is like shouting "Here I am!" into a dark forest filled with hunters. The Great Silence is the sound of wise civilizations staying alive. The only civilizations that make noise are the ones that are too young and foolish to know any better, and they are silenced shortly after they begin to shout.

As children when we hear a bump in the night, natural instinct guides us to hide under the covers while trying not to make a sound. Perhaps as adults we should exercise the same mentality. If the Dark Forest theory is correct, we should not be trying to contact other civilizations.

We should be praying they don't notice us. The paradox is solved: the universe is full of life, but it is also full of fear.

13

Chapter 13: The Zoo & Incubation Hypotheses

The Dark Forest theory paints a grim picture of a cosmos governed by fear and preemptive destruction. It assumes that if advanced civilizations exist, their behavior toward emerging species like ours would be, at best, lethally cautious. But what if this assumption is wrong? What if the psychology of a truly ancient and advanced civilization transcends simple, predatory logic?

This brings us to a class of theories that are philosophically the polar opposite of the Dark Forest. These hypotheses agree that the galaxy is likely populated by powerful, intelligent species, but argue that their silence is not a sign of fear, but of wisdom, ethics, and deliberate non-interference. They are not hiding *from* us; they are hiding themselves *for* us.

The Zoo Hypothesis: A Policy of Non-Interference

The Zoo Hypothesis was first formally proposed by MIT radio astronomer John Ball in 1973. It presents a simple, elegant, and strangely plausible solution to the Fermi Paradox: advanced civilizations exist, they are aware of our existence, but they have deliberately chosen not to make contact. They are treating Earth as a wildlife preserve or a zoo. This hypothesis follows very similar doctrine to how we humans treat other animals on Earth. Most zookeepers and biologists will tell you that the best way to help our wildlife is to keep our presence minimal and try not to interfere. In other words: just leave them alone! Why should the cosmos be any different?

Imagine a nature documentary crew venturing deep into the Amazon rainforest to study a newly discovered, uncontacted tribe. Their prime directive is to observe and document, not to interfere. They would not land a helicopter in the middle of the village and hand out smartphones and antibiotics. To do so would be to irrevocably contaminate and destroy the very culture they wish to study. The introduction of advanced technology, new ideas, and alien biology would shatter the tribe's social structure, religion, and way of life. The well-intentioned act of contact would be, in effect, an act of cultural annihilation.

The Zoo Hypothesis extrapolates this principle to a cosmic scale. A truly advanced civilization, having perhaps witnessed the development and collapse of countless younger species, would understand the catastrophic consequences of premature contact. They would recognize that our civilization, with its warring nations, fragile ecosystems, and burgeoning technologies, is in a delicate, adolescent phase.

Making their presence known would trigger chaos.

- **Social & Political Upheaval:** The confirmation that we are not alone, and are vastly inferior technologically, would shatter our religious, philosophical, and political systems. It could unite humanity, but it is just as likely to cause global panic, new forms of conflict, and the collapse of established power structures.
- **Technological Contamination:** The introduction of advanced technology could be devastating. What would a society that can barely manage nuclear fission do with the secrets to faster-than-light travel or zero-point energy? The power would likely be weaponized, leading to unprecedented destruction.
- **Biological Contamination:** Even a simple exchange of microorganisms could be catastrophic, introducing alien plagues to our biosphere or terrestrial microbes to theirs.

Faced with these risks, the most ethical and logical course of action for a benevolent, advanced civilization would be to enforce a strict, hands-off policy. They would quarantine Earth, observing us from a safe and undetectable distance, much as we observe animals in a zoo or a nature preserve. They are the zookeepers, and we are the exotic, and perhaps endangered, inhabitants. The Great Silence, in this view, is the sound of the zookeepers respecting the "Do Not Feed the Animals" sign. They are waiting for us to mature, to develop into a species that can handle the profound ontological shock of contact.

The Incubation Theory: Waiting for Graduation Day

The Incubation Theory is a close cousin of the Zoo Hypothesis but with a more goal-oriented twist. It also posits that advanced civilizations are deliberately avoiding contact, but it adds a specific condition for when that silence will be broken. In this model, humanity is not just being observed; it is being allowed to "incubate" until it reaches a certain critical milestone of technological and social maturity.

This is the "Prime Directive" from *Star Trek* lore brought to life. In the show, the Federation is forbidden from making contact with any civilization that has not yet developed warp drive (faster-than-light travel). This technological achievement serves as a proxy for a certain level of scientific and social development, marking the species as ready to join the galactic community.

What might our own "graduation day" look like? The criteria could be:

- **Technological:** The development of interstellar travel, the successful harnessing of fusion power, or the creation of a stable artificial general intelligence.
- **Social/Political:** The formation of a stable world government, the complete abolition of warfare, or the successful reversal of planetary-scale environmental damage (like climate change).

- **Ethical:** A demonstrated capacity for long-term empathy, sustainable global management, and a commitment to preserving life.

Once we meet these criteria, the quarantine would be lifted. The "zookeepers" would reveal themselves, not as gods or conquerors, but as peers, welcoming a newly matured species into the cosmic neighborhood.

This theory is inherently optimistic. It transforms the empty-seeming universe into a place filled with patient, watchful guardians. The Great Silence is not a sign of our loneliness, but a testament to our potential. It is the quiet of the examination room, where humanity is in the midst of a final, crucial test. We are being given the chance to solve our own problems, to overcome our own demons, and to prove that we are worthy of a seat at the galactic table. If we succeed, the silence will end. If we fail, the watchers will simply note the unfortunate outcome and continue to wait for the next species to begin its own incubation.

14

Chapter 14: The Simulation Theory

We have explored possibilities ranging from the biological to the sociological, from a rare Earth to a fearful cosmos. We now venture into a realm that is far more radical, a solution to the Fermi Paradox that challenges not just our understanding of life or society, but the very nature of reality itself. This is the Simulation Theory.

The hypothesis, in its modern form, was most famously articulated by Oxford University philosopher Dr. Nick Bostrom in his 2003 paper, "Are You Living in a Computer Simulation?" It proposes that our reality—the stars, the planets, our bodies, and our minds—is not a physical, base reality. Instead, it is an incredibly sophisticated computer simulation programmed by a far more advanced intelligence. If this is true, the answer to the Fermi Paradox could be brutally simple: the reason we don't see any aliens is because the programmers never wrote any into our simulation.

This idea, which sounds like the plot of a science fiction movie like *The Matrix*, is built on a surprisingly logical, three-pronged argument known as Bostrom's Trilemma.

Bostrom's Trilemma

Bostrom does not claim that we *are* living in a simulation. Instead, he argues that one of the following three propositions must be true:

1. **The "Doom" Proposition:** "The fraction of human-level civilizations that reach a 'posthuman' stage is very close to zero."
   A "posthuman" stage is a point of technological maturity where a civilization possesses computing power so immense that it could run vast numbers of high-fidelity simulations of its ancestors or other historical realities. This proposition is essentially a restatement of the Great Filter hypothesis. It suggests that civilizations almost always destroy themselves or otherwise go extinct before they can develop the god-like computing power needed to run such simulations. If this is true, then very few, if any, simulations are ever created.
2. **The "Disinterest" Proposition:** "The fraction of posthuman civilizations that are interested in running ancestor-simulations is very close to zero."
   This proposition assumes that civilizations *do* reach a posthuman stage, but for some reason, they choose not to run simulations. Perhaps they are bound by universal ethical laws that forbid creating conscious, simulated beings. Perhaps they have evolved beyond any interest in their own history or science, and are occupied with pursuits we cannot even imagine. Whatever the reason, if this is true, then even though civilizations *could* run simulations, they don't.
3. **The "Simulation" Proposition:** "The fraction of all people with our kind of experiences that are living in a simulation is very close to one."
   This is the mind-bending conclusion. If the first two propositions are false—meaning that civilizations *do* frequently reach a posthuman stage (we avoid doom) AND they *are* interested in running simulations—then a staggering number of simulations will be created. A single posthuman civilization could run billions of simulations of its past, each containing trillions of simulated conscious beings. The logical consequence is that the number of simulated realities would vastly outnumber the one, single "base

reality." Therefore, any randomly selected conscious being (like you) is overwhelmingly more likely to be one of the trillions of simulated beings than one of the relatively few beings in the original, real world. Statistically, the odds would be that you are in a simulation.

Solving the Great Silence

If we accept that proposition (3) is the true one, the Fermi Paradox dissolves instantly. The silence of our universe is not a deep cosmic mystery, but a simple choice made by the simulation's programmers.

Why would they leave out aliens? The reasons could be numerous:

- **Computational Cost:** Simulating an entire universe is resource-intensive. Simulating a second, independent technological civilization, with its own complex history, biology, and culture, might increase the computational load exponentially. The programmers might have decided it was simply more efficient to focus the simulation's resources on one emerging civilization: ours. Why render a billion complex alien societies if the point of the simulation is to study the development of humanity?
- **A Controlled Experiment:** Our universe could be a scientific experiment. The programmers might be trying to determine how often life arises and evolves in a "pristine" universe, without outside interference. Introducing aliens would contaminate the experiment. The silence is a necessary condition for the validity of their results.
- **The "Solipsistic" Simulation:** The simulation might not even include a full universe. It might only render what is necessary for our perception. The distant galaxies we see through the Hubble telescope might be nothing more than a beautiful, high-resolution backdrop, like the mountains in a video game that you can see but never reach. The universe beyond our solar system

might not be simulated in any detail at all, which would elegantly explain why it appears so empty.

"Glitches in the Matrix"

This theory also provides a novel, if highly speculative, framework for explaining a wide range of otherwise inexplicable phenomena. Anyone who has played a video game has experienced bugs or glitches—characters appearing where they shouldn't, objects behaving unnaturally, or parts of the world failing to load correctly. If our reality is a simulation, why wouldn't it have similar coding errors?

This line of thinking, while purely philosophical, is an entertaining exercise.

- That "UFO" or "UAP" seen performing physics-defying maneuvers? Perhaps it was not an alien craft, but a future aerial vehicle from a different simulation layer that accidentally "glitched" into our reality's sky for a few moments before the error was corrected by the system.
- The Loch Ness Monster? Maybe it was a momentary rendering error, where an asset from a prehistoric simulation layer—a plesiosaur—was accidentally loaded into a modern Scottish lake long enough for a few people to spot it.
- Reports of ghosts, bigfoot, or other paranormal encounters? These could be interpreted as artifacts of the simulation's code—lingering data from a deceased character's file, or a misplaced creature asset from a different part of the program.

While this is fun to ponder, it is important to remember that it is unfalsifiable and therefore not strictly scientific. However, the core of Bostrom's argument is a serious logical proposition. If we don't believe we are in a simulation, we may be forced to conclude that civilizations either almost never reach a posthuman stage (we are likely doomed), or they reach it but lose all interest in their own history (a strange sociological assumption).

The Simulation Theory is the ultimate get-out-of-jail-free card for the Fermi Paradox. It explains the silence perfectly, but at the cost of our entire conception of reality. It postulates that the universe is not silent because it is empty, but because it was designed that way. In essence, it just buries the question of other beings one level deeper. If we're inside a simulation, than who made it? From there we can reapply everything we've discussed so far, just under the additional step of assuming we are inside of a simulation built on some other world.

15

Chapter 15: Interdimensional & String Theory

The solutions to the Fermi Paradox we've discussed so far, even the radical Simulation Theory, largely operate within the familiar confines of our four-dimensional universe (three spatial dimensions and one of time). They assume that if aliens exist, they exist *out there*, among the same stars and galaxies we see. But what if the stage itself is far larger and more complex than we perceive? What if other beings exist not on distant planets, but in different realities, different dimensions, operating under physical laws that are subtly or radically different from our own?

This line of inquiry takes us to the cutting edge of theoretical physics, to concepts like string theory and the multiverse. It proposes a solution to the Great Silence that is both profound and mind-altering: the universe isn't silent, we are simply tuned to the wrong channel.

The "Radio" Analogy

To grasp this concept, let us fully develop an analogy from your original notes. Imagine a person who lived and died in the year 1700. Their world was filled with the tangible sounds of nature, animals, and human activity. If you were to tell them that the very air around them was filled with invisible music, complex conversations, and moving pictures, they would think you were a mystic or a madman. They lack the conceptual framework to understand electromagnetism and the technology—a radio or television—to perceive it. Yet, the radio waves are there, passing through their body and their home, completely undetected.

The Interdimensional Hypothesis suggests that we may be in a similar state of ignorance. We are "deaf" to other forms of existence that may be all around us, separated not by physical distance, but by a "frequency" of reality that our senses and scientific instruments are not designed to detect. What we perceive as the Great Silence might simply be the result of a profound sensory and technological limitation. The "aliens" are not on a distant star; they are in the room with us, but in the "next channel over" on the cosmic dial.

String Theory and Hidden Dimensions

For centuries, this idea was pure philosophy. But in the late 20th century, theoretical physics began to provide a potential mathematical foundation for it. This foundation is String Theory.

At its most basic level, string theory challenges our understanding of the fundamental particles that make up our universe. For decades, physics has operated on the "point-particle" model, viewing electrons, quarks, and photons as dimensionless points. String theory replaces these points with something new: a tiny, one-dimensional, vibrating loop, or "string."

This seemingly simple replacement has monumental consequences. According to the theory, every particle in the universe is just the same kind of string, but vibrating at a different frequency. The frequency of the string's vibration determines the particle's properties—its mass, its charge, its spin. An electron is a string vibrating in one pattern; a photon is the same string vibrating in another. As Columbia University physicist Dr. Brian Greene describes it, this adds "the new microscopic layer of a vibrating loop to the previously known progression from atoms through protons, neutrons, electrons, and quarks."

This elegantly unifies the particle zoo into a single, underlying entity. But for the mathematics of string theory to work, it requires something extraordinary: the universe must have more than the three spatial dimensions we experience. In fact, the leading version of the theory, M-theory, requires a total of ten spatial dimensions.

Where are these other six dimensions? String theorists believe they are "curled up" or "compactified" to a size so infinitesimally small that we cannot perceive them. Imagine a garden hose. From a great distance, it appears to be a one-dimensional line. But up close, you can see it has a second dimension: the circular circumference you could travel around. In the same way, every point in our familiar three-dimensional space may, in fact, be a tiny, six-dimensional geometric shape, but on a scale far too small for us to ever directly access.

The "Brane-World" Scenario

This leads to an even more dramatic concept: the "brane-world" scenario. In this model, our entire three-dimensional universe may exist as a membrane, or "brane," floating within a higher-dimensional space, often called the "bulk." There could be other, parallel branes—other universes—floating alongside our own, perhaps only a microscopic distance away in a higher dimension.

Most of the forces and particles of our universe are confined to our brane. The photons that make up light, the electrons in our atoms—they are all stuck here. This is why we cannot see these other universes. However, string theory suggests that one force might be an exception: gravity. Gravity, in some models, might be able to "leak" between branes.

The implications are staggering. It means that other universes, with potentially different physical laws and different forms of "life," could be existing parallel to our own. These universes are not billions of light-years away; they are, in a sense, right next to us, but inaccessible.

A New Answer to the Paradox

This provides a radical new solution to the Fermi Paradox. The reason we don't see aliens from other planets is because life doesn't need to evolve on other planets. It could evolve in these parallel branes, in these other dimensions. The Great Silence is not an absence of life in the cosmos, but our inability to perceive life that exists outside our own three-dimensional brane.

This could also offer a framework for understanding historical myths and legends. Could the "realms" of Norse mythology (like Yggdrasil, the world tree connecting nine realms), the "planes" of existence in Hindu and Buddhist cosmology, or the descriptions of angels, djinn, and fairies be folk interpretations of beings from another brane briefly intersecting with our own? A being that can manipulate extra dimensions would appear to us as a god or a spirit, capable of appearing and disappearing at will, passing through solid objects, and defying our known laws of physics.

The Interdimensional Hypothesis is at the very edge of scientific thought. It is speculative, untestable with current technology, and borders on the metaphysical. Yet, it is built on the most advanced mathematics our species has ever produced. It suggests that our search for life among the stars may be like the man who has lost his keys searching only under the lamppost, not because that's where he lost them, but because that's where the light is. The rest of the universe, silent and dark to us, may be ablaze with life and intelligence, waiting for us to figure out how to build a "radio" that can tune in to the other dimensions.

16

Chapter 16: Other Notable Theories

The grand, sweeping theories we've explored—from Rare Earth to the Dark Forest to Simulation Theory—provide comprehensive, if unsettling, explanations for the Great Silence. But the intellectual landscape of the Fermi Paradox is vast, populated by dozens of other hypotheses. Some are simple and practical, while others are as imaginative as any science fiction. This chapter serves as a tour of some of these other notable possibilities.

Each of these theories tackles a specific aspect of the Drake Equation, arguing that we have overestimated a particular variable or failed to consider a crucial factor. They may not be as all-encompassing as the Great Filter, but they provide valuable, focused perspectives on why the cosmos might appear so quiet.

1. The Early Bird Theory (We Are First)

This is perhaps the simplest and most optimistic solution to the paradox. It accepts all the optimistic assumptions about the prevalence of life and intelligence but offers a bold conclusion: we are, by sheer cosmic chance, one of the very first technological civilizations to arise.

The universe may be 13.8 billion years old, but the conditions for life have not always been so favorable. The early universe was filled with hot, short-lived Population III stars composed almost entirely of hydrogen and helium. It took generations of stars to live and die, forging heavier elements ("metals") in their cores and scattering them through supernovae, before there were enough raw materials to form rocky planets like Earth.

Furthermore, the early universe was a much more violent place, with more frequent galaxy collisions, quasar activity, and sterilizing gamma-ray bursts. The argument, championed by astronomers like Avi Loeb, is that the "cosmic habitable age" may have only begun relatively recently. Our sun, at 4.6 billion years old, formed well into this calmer, metal-rich era.

If this is the case, then we are not late to the party; we are the ones setting it up. There are no ancient, million-year-old civilizations to find because the timeline hasn't allowed for them. Other civilizations may be emerging concurrently with us, or they may be millions of years behind us, still in their microbial or pre-industrial stages. The Great Silence is simply the sound of an empty stage before the main performance has begun. We are the "early birds," and the galaxy is our unclaimed inheritance.

2. Signal Propagation Issues (The Message is Lost)

This practical theory suggests that advanced civilizations may well be out there and may even be actively broadcasting, but we are simply unable to receive their signals. There are several reasons this might be the case.

- **The Signal is Too Weak:** Space is vast. An electromagnetic signal's strength decreases exponentially with distance (an inverse-square law). Even a powerful signal directed at us from a distant star might attenuate so much over its journey of hundreds or thousands of light-years that by the time it reaches Earth, it is indistinguishable from the background cosmic noise. We may be bathed in alien signals that are simply too faint for our current radio telescopes to detect.

- **We're Listening to the Wrong "Channel":** We assume that radio waves are the logical medium for interstellar communication, but this could be a profound failure of imagination. A civilization just a thousand years more advanced than us might have discovered a form of communication that is far more efficient and instantaneous—perhaps using manipulated neutrinos, gravitational waves, or some form of physics we haven't even conceived of yet. They could be shouting, but we, with our primitive radio "ears," are deaf to it. It's like trying to listen for a Wi-Fi signal with an AM radio.
- **The Signal is Too Efficient:** Our own communication methods have become progressively more efficient and less "leaky." Early television and radio broadcasts were omnidirectional, blasting signals out into space. Modern communication uses fiber optics, targeted cell signals, and highly compressed data streams. An advanced civilization would likely be even more efficient, wasting no energy broadcasting into empty space. Their communications would be tightly-beamed, encrypted, and compressed. Trying to intercept their signals would be like trying to read a single fiber-optic cable from a light-year away.

3. Difficulties of Galactic Colonization (The Physics is Too Hard)

Many formulations of the Fermi Paradox assume that a sufficiently advanced civilization would inevitably colonize the galaxy. But what if that assumption is wrong? The physical barriers to interstellar travel may be far more formidable than we imagine.

- **The Tyranny of Lightspeed:** Einstein's theory of relativity posits that the speed of light is an absolute, unbreakable speed limit. Even approaching this speed requires colossal amounts of energy. A journey to the nearest star system, Proxima Centauri, would take over four years at the speed of light. A journey to the other side of our galaxy would take 100,000 years.

- **The Dangers of Travel:** A spaceship traveling at a significant fraction of lightspeed would be subject to immense dangers. A collision with even a tiny, pebble-sized dust grain would be a catastrophic, explosive event. The ship would also need to be heavily shielded against cosmic rays and radiation.
- **The Problem with Wormholes and Warp Drive:** While these are staples of science fiction, they may be physical impossibilities. Wormholes, if they can exist, might be microscopically small and catastrophically unstable. Warp drive may require "exotic matter" with negative mass, which may not exist.

It could be that every civilization, no matter how advanced, eventually runs up against the same hard physical laws. Interstellar colonization is simply too difficult, too expensive, or too dangerous to be a viable strategy. Civilizations may live and die in their home solar systems, forever isolated by the vast, impassable gulfs between the stars.

4. Different Definitions of Life (We Don't Recognize Them)

Our entire search is predicated on "life as we know it"—carbon-based, water-dependent, DNA-coded organisms. This is a reasonable starting point, but it could also be a critical limitation. Life elsewhere could be so radically different that we wouldn't even recognize it as life.

Imagine silicon-based life forms on a world with liquid methane oceans. Or consider life as a vast, sentient plasma cloud in a nebula, or a crystalline intelligence that "lives" on geological timescales. Perhaps life has evolved to become non-biological, uploading its consciousness into a distributed computational network in the vacuum of space. Such beings might be all around us, but we, with our biological biases, are completely blind to their existence. We are searching for familiar patterns in a universe that may be filled with a kind of life so alien it is, to our senses, indistinguishable from inert matter or natural phenomena.

5. Cost-Benefit Ratio (We're Not Worth the Effort)

This is the most humbling, and perhaps the most cynical, of the theories. It suggests that advanced civilizations may be aware of us, have the capability to contact us, but have simply chosen not to because it is not worth their time or resources.

A civilization that has mastered interstellar travel may have already encountered thousands of "primitive" worlds like ours. We may simply not be that interesting. They may have determined that the immense cost of mounting a mission or a communication project to a backwater solar system like ours far outweighs any potential benefit—be it scientific, cultural, or material.

Think of it from a human perspective. You are aware that there are uncontacted ant colonies in the Amazon rainforest. You have the technological means to go there and try to communicate with them. But would you? It is an enormous effort for what would likely be very little gain. To a hyper-advanced galactic civilization, humanity might be nothing more than a curious but ultimately insignificant ant colony. The silence, in this case, is not malice or fear; it's simple indifference.

6. Societal Apex Theory (We're Not Important Enough)

There may be something to say for the crowd who believes that only high ranking government officials have the pleasure of conversing with extraterrestrials. It makes sense. If any highly-advanced intelligent biological entity from beyond our planet's biosphere wanted to contact humanity, there are multiple reasons why they would not approach a common person, but rather identify leadership and make a more structured first contact. It may be the case that the average human hasn't encountered any evidence of extraterrestrial biological entities, because such entities' diplomats are preferentially contacting only humanity's most elite and highest-ranked leaders.

The primary reason for such exclusivity would likely boil down to a simple, universal principle: efficiency. An advanced civilization, capable of traversing interstellar distances, would operate on a level of logistical and strategic thinking far beyond our own. Attempting to communicate with eight billion individuals simultaneously would be chaotic, pointless, and informationally overwhelming. Instead, they would logically seek out the central nodes of our planet's power and communication structures. Just as a human ambassador doesn't negotiate a treaty by polling every citizen of another country, an alien delegation would engage with the individuals who hold the levers of control—heads of state, military leaders, and heads of global coalitions like the United Nations. Meeting with these leaders is not an act of elitism, but the most direct and efficient path to achieving any objective, whether it be a warning, a trade proposal, or a simple announcement of their presence.

Beyond simple efficiency, any truly intelligent species would recognize the monumental psychological shock that first contact would inflict upon humanity. The revelation that we are not alone could shatter foundational religious, social, and political frameworks, potentially leading to mass panic, existential crises, and widespread chaos. To mitigate this, the aliens might see our leaders as crucial "gatekeepers of stability." By briefing a small, powerful group first, they allow the planet's established authorities to control the narrative, manage the release of information, and deploy resources (like military or emergency services) to prevent societal collapse. In this view, an exclusive meeting is not a snub to the general populace, but a necessary quarantine measure—a controlled burn designed to prevent a global wildfire of fear and disorder.

Furthermore, the choice of who to meet would be deeply symbolic, a reflection of what the aliens deem most important or representative about our species. Meeting with a political leader like a president or the UN Secretary-General would signify an engagement with our capacity for self-governance and our political structure. However, a civilization interested in our intellectual potential might bypass politics entirely and request a meeting with our most brilliant scientific minds—a council of Nobel laureates or the heads of institutions like SETI. Conversely, a species concerned with ethics or consciousness might seek out spiritual figures like the Pope or the Dalai Lama. In this sense, the choice of contact is the ultimate first impression; it is a message in itself, telling us more about the aliens' own values and their reason for visiting than any words they might initially speak.

Pop culture already acknowledges societal apex theory as a realistic possibility. Countless films portray scenes of aliens landing on the White House lawn, meeting in secret with world leaders, or attacking high-profile military sites. This type of contact naturally feels more realistic than an extraterrestrial biological entity traveling across the universe just to impact a random person's day. When we think of aliens, people often quote the typical, "Take me to your leader" line, but we seldom hear anyone impersonating aliens saying, "Take me to a random, inconsequential person". Whether such aliens would be acting out of care for humanity's interests or something far more sinister, the preferential selection of society's apex members would likely remain the same.

Part IV: The Future and The Silence

Our journey through the Fermi Paradox has taken us from the microscopic definition of life to the mind-bending frontiers of theoretical physics. We have navigated the probabilities of the Drake Equation and explored a labyrinth of potential solutions, from the lonely rarity of Earth to the terrifying logic of a cosmic dark forest. We are left with a universe of profound possibilities, but no certain answers. The Great Silence remains.

So, where do we go from here? Do we abandon the search, resigned to our cosmic solitude? Or do we push forward, refining our tools and our thinking in the hope that we might one day hear a faint whisper from the void?

In this final section, we will look to the future. We will examine the next generation of scientific instruments that are poised to revolutionize our search, and we will conclude by reflecting on what the Great Silence means for humanity's own future, regardless of whether we are alone or not.

17

Chapter 17: The New Search

For its entire history, the Search for Extraterrestrial Intelligence (SETI) has been a passive, listening exercise. We have been like someone standing by the shore of a vast ocean, waiting for a message in a bottle to wash up. The search has also been chronically underfunded, relying on limited access to telescopes and private donations. But we are now entering a new era of exploration, equipped with tools of unprecedented power and scope. The new search is proactive, multi-faceted, and far more ambitious than anything that has come before.

The James Webb Space Telescope (JWST): Sniffing for Atmospheres

Launched on Christmas Day in 2021, the James Webb Space Telescope is the most powerful and complex space observatory ever built. While it is not designed to detect radio signals, it has a capability that may be even more profound: it can analyze the chemical composition of exoplanet atmospheres.

As an exoplanet transits in front of its host star, a tiny fraction of the starlight passes through the planet's atmosphere. By capturing this light with its massive mirror and splitting it into a spectrum, JWST can identify the chemical fingerprints—the absorption lines—of the gases present in that atmosphere. This is called transmission spectroscopy.

This allows us to search for "biosignatures." These are gases that are strongly associated with biological processes and are unlikely to exist in large quantities on their own. On Earth, the most obvious biosignature is the abundant free oxygen in our atmosphere, which is almost entirely produced and maintained by photosynthetic life.

JWST will be able to search the atmospheres of nearby, Earth-sized planets for tell-tale signs of life:

- **Oxygen (O2) and Methane (CH4):** Finding both of these gases in the same atmosphere would be a very strong indicator of life. Methane and oxygen are chemically reactive and would quickly destroy each other. To find them in a stable equilibrium suggests that something—like the microbes in a cow's stomach producing methane and the Amazon rainforest producing oxygen—is constantly replenishing them.
- **Water Vapor (H2O):** Confirming the presence of water is a key step in assessing habitability.
- **Other potential biomarkers,** such as nitrous oxide or even industrial pollutants like chlorofluorocarbons (CFCs), which would be an unambiguous "technosignature."

For the first time, we have the ability to move beyond simply finding habitable-zone planets to actually analyzing their chemistry. A confirmed detection of a biological atmosphere on another world would be a monumental discovery, proving that fl (the fraction of planets where life arises) is not a vanishingly small number. We would know that life is not unique to Earth, even if we hadn't yet found intelligence.

Breakthrough Listen: The Ultimate Radio Search

While JWST looks for chemistry, the SETI search for signals continues, but on a scale that dwarfs all previous efforts. In 2015, billionaire Yuri Milner announced the creation of Breakthrough Listen, the most comprehensive and intensive search for extraterrestrial communications ever undertaken.

With $100 million in funding over ten years, Breakthrough Listen is a systematic survey of the sky, far surpassing the scope of all previous SETI projects combined. It is:

- Surveying 1,000,000 of the nearest stars to Earth.
- Listening to the plane and center of our own Milky Way galaxy.
- Scanning 100 of the nearest galaxies beyond our own.
- Covering a massive radio band (from 1 to 10 GHz), far more than the narrow hydrogen line targeted by early searches.
- Using thousands of hours of observation time on two of the world's most powerful radio telescopes: the Green Bank Telescope in West Virginia and the Parkes Observatory in Australia.

The sheer volume of data being collected is unprecedented. The project is "listening" to billions of channels simultaneously. This massive increase in coverage dramatically improves the chances of stumbling upon a signal, if one is there to be found.

Looking for Lasers and Artifacts

The new search is also expanding beyond the radio spectrum. If a civilization wanted to send a deliberate, high-bandwidth message, they might use powerful optical lasers instead of radio. These laser signals would be brief but incredibly bright flashes of monochromatic light. Projects are now underway to scan the sky for such artificial signals.

Furthermore, some are proposing a search for physical artifacts within our own solar system. If an advanced civilization has ever explored our region of the galaxy, they may have left behind a probe—a "lurker"—to observe us. These could be small, perhaps only a few meters in size, and could be hiding in stable orbits, such as in the asteroid belt or at Lagrange points. Astronomers like Avi Loeb are spearheading efforts like the Galileo Project to search for such objects, arguing that we should be actively looking for extraterrestrial technology within our own backyard, not just listening for distant signals. One common example of such an artifact is the famous Black Knight satellite, commonly discussed on conspiracy forms and scientific stages alike.

The Black Knight satellite is the subject of a long-standing conspiracy theory that claims an artificial satellite of extraterrestrial origin has been orbiting Earth for approximately 13,000 years. Proponents of the theory point to a collection of unrelated stories as evidence. These include supposed alien radio signals detected by Nikola Tesla in 1899, long-delayed radio echoes heard in the 1920s, and unattributed satellite discoveries in the 1950s, before the launch of Sputnik. The name "Black Knight" itself is thought to have originated from a British rocket program and was later retroactively applied to this collection of mysteries.

A key piece of "evidence" for believers is a set of photographs taken during the 1998 STS-88 space shuttle mission, which show a dark, irregularly shaped object. However, NASA has identified this object as a thermal blanket that was lost by astronauts during a spacewalk. Scientists and skeptics maintain that each element of the Black Knight theory has a logical and terrestrial explanation. The signals Tesla detected are now believed to have been from pulsars, and the long-delayed echoes are a known radio phenomenon. Despite the lack of credible evidence, the Black Knight satellite remains a popular topic in conspiracy discussions and online forums, fueled by a fascination with the unknown and the possibility of extraterrestrial life.

The new search is a paradigm shift. We are moving from passively waiting to proactively scanning, from listening on one channel to observing across the spectrum, and from looking only at distant stars to searching our own cosmic neighborhood. If there is anything to be found, our chances of finding it in the coming decades are greater than they have ever been before.

Conclusion: Humanity and The Great Silence

We stand at the shore of the cosmic ocean. We have cast our nets of inquiry into its depths, exploring the probabilities of life, the nature of intelligence, and the vast, silent expanse of time. We have weighed the unsettling logic of the Dark Forest against the hopeful quarantine of the Zoo. We have questioned the very fabric of our reality and considered our own lonely rarity in a sterile cosmos. And yet, after all our searching, all our theories, all our speculation, we are left where we began: with the profound and unanswered question of Enrico Fermi. Where is everybody?

The Great Silence is, for now, an empirical fact. The heavens do not speak to us. We have found no signals, no artifacts, no unambiguous signs of intelligence beyond our own world. We may never have a definitive answer to the Fermi Paradox in our lifetimes. The solution may be a complex mix of the hypotheses we've discussed, or it may be something so utterly strange that we have not yet conceived of it.

But the value of the Fermi Paradox is not contingent on finding a solution. Its true power lies in its function as a great cosmic mirror. In asking "Where are they?" we are forced, inevitably, to ask "Who are we?" Each proposed solution to the paradox reflects our own civilization's deepest anxieties, its highest aspirations, and its most critical vulnerabilities.

If we fear the Great Filter is ahead of us, we are acknowledging the precariousness of our own technological adolescence. We see the specter of nuclear war, climate change, and uncontrolled AI not just as terrestrial problems, but as cosmic hurdles that may have doomed countless civilizations before us. The silence becomes a dire warning, a call to action to get our own house in order, to prove that we can be the species that finally overcomes the odds.

If we are drawn to the Dark Forest theory, we are projecting our own history of conflict, colonization, and fear onto the canvas of the cosmos. We see in the silence the logical endpoint of unchecked suspicion and existential competition. It forces us to question whether contact is even a desirable goal, and to recognize that survival may depend more on wisdom and restraint than on bold exploration.

If we find comfort in the Zoo Hypothesis, we are expressing a deep-seated hope for cosmic benevolence and a desire for mentorship. We are casting ourselves as the promising but unruly children in a galaxy of wise elders. The silence becomes a challenge to mature, to achieve a global society worthy of joining a galactic community. It is an expression of faith in our own potential.

And if we believe we are alone, a consequence of a Rare Earth or a singular genesis event, the silence transforms into something else entirely. It becomes a statement of profound responsibility. If our world is the only bastion of complex life in the galaxy, if our consciousness is the only consciousness capable of observing and understanding the universe, then we are the custodians of something infinitely precious. The destruction of our civilization would not just be a terrestrial tragedy; it would be a cosmic one. It would be the extinguishing of the universe's only known light. In this view, the silence imbues our existence with a meaning and a duty of the highest order. We are the inheritors of an empty cosmos, and what we do with that inheritance matters.

The search for extraterrestrial intelligence is more than just a scientific quest; it is a deeply humanistic one. It is the ultimate expression of our species' innate curiosity, our desire to connect, and our need to understand our own place in the grand scheme of things.

The journey to understand the Great Silence is, in the end, a journey to understand ourselves. We may find, after decades more of searching, that the silence was simply a failure of our instruments, and that the cosmos is indeed a bustling, vibrant place. Or we may find that the silence is as deep and as absolute as it appears, and that we are, for all practical purposes, alone.

Either possibility is, in its own way, breathtaking. Whether we are one voice in a cosmic choir or a single, solitary voice in an empty cathedral, our destiny is the same. We must continue to listen, to explore, to learn, and to build a civilization that is stable, wise, and resilient. We must build a civilization worthy of surviving the Great Silence. For in the end, the most important answer we seek is not whether there is intelligence out there, but whether there will continue to be intelligence right here. The search continues.

Appendix: A Deeper Dive into the Drake Equation

Throughout this book, I have treated the Drake Equation as a conceptual framework for organizing our thinking about the Fermi Paradox. However, it is also a formula that invites us to plug in numbers, to see how different assumptions can lead to wildly divergent outcomes. This appendix presents a comparison of three distinct approaches to the equation: the original optimistic estimates from the 1961 Green Bank meeting, a modern pessimistic view informed by hypotheses like Rare Earth and the Great Filter, and a modern optimistic view informed by discoveries in exoplanet science and theories of cosmic abundance. The Drake Equation, conceived by Dr. Frank Drake in 1961, is not a physical law but a probabilistic framework designed to organize the factors that determine the number of detectable civilizations in our galaxy. It serves as a roadmap for the search for extraterrestrial intelligence.

The equation is written as:

$$N = R_{*} \times f_p \times n_e \times f_l \times f_i \times f_c \times L$$

This appendix provides a more detailed examination of each of these seven variables, reflecting our current scientific understanding.

R* – The Rate of Star Formation

This variable represents the average rate at which new, stable stars are formed in the Milky Way galaxy each year. These stars are the potential hosts for planetary systems. While early estimates were rough, modern astronomy, using data from space telescopes that can peer into stellar nurseries like the Orion Nebula, has provided a much clearer picture.

The current consensus is that the Milky Way converts about 1.5 to 3 solar masses of gas and dust into new stars annually. Because many stars are smaller than our sun, this translates to an average of **three to five new stars** igniting each year. This steady, relentless process over 13 billion years has created the hundreds of billions of stars in our galaxy. The raw potential for new solar systems is, and always has been, abundant.

fp – The Fraction of Stars with Planets

This variable asks what percentage of the stars formed actually host planetary systems. For most of history, we had no data to answer this. Dr. Drake's original estimate of 50% was a hopeful guess.

The last three decades have revolutionized this term. Using the "wobble" (radial velocity) and "transit" (brightness dip) methods, missions like NASA's Kepler Space Telescope have discovered thousands of exoplanets. The results have been staggering. The data indicates that planets are not an exception but the universal rule.

The current scientific estimate for **fp is approximately 1**, meaning that, on average, there is at least one planet for every star in the galaxy. The Milky Way is home to hundreds of billions, if not trillions, of worlds.

ne – The Number of Habitable Worlds

This variable narrows the focus from all planets to only those that are environmentally suitable for life as we know it. This is the search for "Goldilocks" planets. The primary requirement is the potential for liquid water, which necessitates that a planet orbits within its star's Circumstellar Habitable Zone (CHZ)—not too hot, not too cold.

However, other factors are critical:

- **Atmosphere:** A planet needs sufficient mass and gravity to hold onto a substantial atmosphere, which is required to maintain liquid water on the surface.
- **Subsurface Oceans:** Moons outside the CHZ, like Europa and Enceladus, may host vast liquid water oceans beneath their icy shells, kept warm by tidal heating from their parent planets. This dramatically expands the potential locations for life.
- **Protective Shields:** A habitable world needs a global magnetic field (generated by a molten core) to deflect sterilizing solar wind and an atmosphere to absorb harmful radiation and burn up meteors.

Despite these stringent requirements, the sheer number of planets is so vast that the final tally is still enormous. Current estimates based on Kepler data suggest there could be as many as **10 to 40 billion potentially habitable, Earth-sized planets** in the Milky Way.

fl – The Fraction Where Life Arises

Here we leave the relative certainty of astronomy and enter the speculation of biology. Of all the habitable worlds, on what fraction does life actually begin?

- **The Optimistic View (fl ≈ 1):** Life is a natural outcome of cosmic chemistry. The building blocks are common, and life appeared on Earth very early, suggesting it is not a difficult step. The universe is hard-wired for biology.
- **The Pessimistic View (fl ≈ 0):** Assembling the first self-replicating cell is a hurdle of staggering, near-miraculous improbability. Earth may be a singular fluke in a sterile galaxy.

We have only one data point: Earth. Until a "second genesis" is found, this remains perhaps the most significant unknown in the equation.

fi – The Fraction Where Intelligence Arises

Given a planet with life, on what fraction does that life evolve to become technologically intelligent?

- **The Optimistic View:** Intelligence is a powerful evolutionary advantage. Convergent evolution suggests that, given enough time, intelligence will inevitably arise to fill that ecological niche.
- **The Pessimistic View:** On Earth, out of billions of species over nearly 4 billion years, only one has developed high technology. This suggests the path to intelligence is incredibly narrow and requires a highly specific set of evolutionary circumstances (large brains, hands, language, etc.). This may be an exceptionally rare outcome.

fc – The Fraction That Become Communicative

Of the species that become intelligent, what fraction develop the technology and the desire to send detectable signals into space?

An intelligent species might never develop technology if, for example, it were aquatic and could not discover fire or metallurgy. Alternatively, a civilization's social, political, or religious systems might discourage scientific inquiry or looking to the stars. While curiosity seems a natural trait of intelligence, it is not guaranteed to lead to radio telescopes.

L – The Longevity of a Communicative Civilization

This final, haunting variable represents the average length of time a civilization survives while being technologically active. It is the great bottleneck. Even if life and intelligence are common, if `L` is short, the chances of two civilizations' technological windows overlapping are vanishingly small.

- **A Short L (A few hundred or thousand years):** This is the core of the Self-Destruction Hypothesis. Civilizations may inevitably destroy themselves through nuclear war, climate change, engineered pandemics, or uncontrolled AI before they can expand into space. If `L` is short, the silence is the sound of countless failed civilizations.
- **A Long L (Millions or billions of years):** A civilization that survives its technological adolescence might become virtually immortal, mastering its energy needs and overcoming all existential threats. If even a few such civilizations exist, their signals should be easily detectable.

The profound uncertainty in `L`, combined with the unknowns of `fl`, `fi`, and `fc`, is what keeps the Drake Equation from providing a definitive answer. It does not solve the Fermi Paradox, but it masterfully illuminates the questions we must continue to ask.

Notes on the Estimates

• **Original Estimate:** This reflects the early, buoyant optimism of the first SETI pioneers. They assumed that once life starts, it is likely to proceed toward intelligence, but that civilizations may not last for an exceptionally long time. A result of 10 detectable civilizations suggested that a dedicated search was a worthwhile, if challenging, endeavor.

• **Modern Pessimistic View:** This viewpoint incorporates our modern understanding that planets are common (fp=1) but assumes that multiple "filters" are severe. It posits that truly Earth-like worlds with stable climates are rare (low ne), the origin of life is a near-miracle (very low fl), the leap to intelligence is an evolutionary fluke (very low fi), and that civilizations that do arise tend to self-destruct very quickly (very low L). The result is a number so infinitesimally small that it predicts we are effectively alone in the galaxy.

• **Modern Optimistic View:** This viewpoint also accepts that planets are common. It takes an optimistic stance on ne by including possibilities like life in subsurface oceans on icy moons. It assumes that life is a cosmic imperative that will arise wherever conditions allow (fl=1). It further assumes that intelligence is a common outcome of evolution (convergent evolution) and that at least some civilizations survive their technological adolescence to become incredibly long-lived (very high L). The result is a galaxy filled with a million active, detectable civilizations, making the Great Silence a profound and deeply puzzling paradox.

This book stands on the shoulders of giants—the scientists, philosophers, and writers who have dedicated their careers to exploring the cosmos and humanity's place within it. The following list is not exhaustive, but it represents the key primary sources and foundational texts that have informed the ideas and theories discussed in "The Great Silence." For any reader wishing to delve deeper into these fascinating topics, these works are the ideal place to start.

Ball, John A. "The Zoo Hypothesis." *Icarus*, vol. 19, no. 3, 1973, pp. 347–349.

Bostrom, Nick. "Are You Living in a Computer Simulation?" *Philosophical Quarterly*, vol. 53, no. 211, 2003, pp. 243-255.

Cocconi, Giuseppe, and Philip Morrison. "Searching for Interstellar Communications." *Nature*, vol. 184, no. 4690, 1959, pp. 844–846.

Drake, Frank. "Project Ozma." *Physics Today*, vol. 14, no. 4, 1961, pp. 40–46.

Greene, Brian. *The Elegant Universe: Superstrings, Hidden Dimensions, and the Quest for the Ultimate Theory*. W. W. Norton & Company, 2003.

Hanson, Robin. "The Great Filter - Are We Almost Past It?" 1998. Online.

Hart, Michael H. "An Explanation for the Absence of Extraterrestrials on Earth." *Quarterly Journal of the Royal Astronomical Society*, vol. 16, 1975, pp. 128–135.

Liu, Cixin. *The Dark Forest.* Translated by Joel Martinsen, Tor Books, 2015.

Miller, Stanley L. "A Production of Amino Acids Under Possible Primitive Earth Conditions." *Science*, vol. 117, no. 3046, 1953, pp. 528–529.

Sagan, Carl, et al. *Murmurs of Earth: The Voyager Interstellar Record*. Random House, 1978.

Ward, Peter D., and Donald Brownlee. *Rare Earth: Why Complex Life Is Uncommon in the Universe*. Copernicus Books, 2000.

A

B

C

D

K

L

M

JOIN THE COMMUNITY - BECOME A PART OF THE STORY

Facebook Page - The Great Silence - A Journey Through the Fermi Paradox

Instagram - @ DoctorHuskyBooks

Reddit - u/ Doctor_Husky

About the Author

Dr. John Prorock is a decorated United States Air Force veteran and distinguished Ph.D. holder. Standing at the intersection of military strategy and cutting- edge space exploration, Dr. Prorock has worked directly with NASA and SpaceX to support the launch of multiple manned space missions, ensuring the safety and success of astronauts traveling into the final frontier. With a career dedicated to the technical and operational realities of the air and space domains, Dr. Prorock brings a unique, high-stakes perspective to the Fermi Paradox. While others approach "The Great Silence" as a philosophical question, he approaches it with the analytical rigour of a mission scientist and the strategic mindset of a military leader. When he isn't supporting the next generation of space exploration, Dr. Prorock focuses on bridge-building between complex scientific concepts and the public, helping us understand our place—and our potential future—among the stars.

www.ingramcontent.com/pod-product-compliance
Lightning Source LLC
LaVergne TN
LVHW090420210926
848480LV00002B/634
9798218911126